n-BODY
PROBLEMS AND MODELS

∩-BODY
PROBLEMS AND MODELS

Donald Greenspan

University of Texas at Arlington, USA

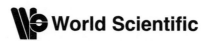

World Scientific

NEW JERSEY • LONDON • SINGAPORE • BEIJING • SHANGHAI • HONG KONG • TAIPEI • CHENNAI

Published by

World Scientific Publishing Co. Pte. Ltd.

5 Toh Tuck Link, Singapore 596224

USA office: Suite 202, 1060 Main Street, River Edge, NJ 07661

UK office: 57 Shelton Street, Covent Garden, London WC2H 9HE

British Library Cataloguing-in-Publication Data
A catalogue record for this book is available from the British Library.

N-BODY PROBLEMS AND MODELS

ISBN 981-238-722-6

Printed in Singapore by World Scientific Printers (S) Pte Ltd

Preface

This book is concerned with computer simulation of scientific and engineering phenomena in a fashion which is consistent with the two principles:

(1) All things change with time, and
(2) All material bodies consist of atoms and/or molecules.

Applications include solitons, crack development, biological sorting, saddle surfaces, rotating tops, bubbles in liquids, liquid surface adhesion, relativistic oscillation and development of turbulent flows. Theoretically we develop discrete equations with conservation laws which are identical to those of continuum mechanics. Our molecular studies are in complete accord with modern nanophysics.

Graduates and professional researchers in mathematics, physics, materials science, fluid dynamics, and electrical and mechanical engineering will find this book a contemporary resource for their work on modelling physical phenomena.

Finally, I wish to thank Ann Kostant, Executive Editor, Mathematics and Physics, Birkhauser, Boston, for permission to use material from my book PARTICLE MODELING (1997).

Donald Greenspan
Arlington, Texas 2004

Contents

Problem Statement

The general N-body problem is usually formulated for $N \geq 2$ as follows. In cgs units and for $i = 1, 2, \ldots, N$, let P_i of mass m_i be at $\vec{r}_i = (x_i, y_i, z_i)$, have velocity $\vec{v}_i = (v_{i,x}, v_{i,y}, v_{i,z})$, and have acceleration $\vec{a}_i = (a_{i,x}, a_{i,y}, a_{i,z})$ at time $t \geq 0$. Let the positive distance between P_i and P_j, $i \neq j$, be $r_{ij} = r_{ji} \neq 0$. Let the force on P_i due to P_j be $\vec{F}_{ij} = \vec{F}_{ij}(r_{ij})$, so that the force depends only on the *distance* between P_i and P_j. Also, assume that the force $\vec{F}_{ji}$ on P_j due to P_i satisfies $\vec{F}_{ji} = -\vec{F}_{ij}$. Then, given the initial positions and velocities of all the $P_i, i = 1, 2, 3, \ldots, N$, the general N-body problem is to determine the motion of the system if each P_i interacts with all or part of the other P_j's in the system.

The prototype N-body problem was formulated around 1900 and was a collisionless problem. In it the P_i were the sun and the then known eight planets and the force on each P_i was gravitational attraction. However, since 1900 and up to the present, a variety of other N-body problems have come to be of interest in the sciences and in engineering. These problems will be categorized according to the choices $N = 1, 2 \leq N \leq 100, 100 < N \leq 10000, 10000 < N$. These categories have been determined in accordance with the capabilities of a Digital Alpha 533 personal scientific computer, which has been used for all the examples to be described.

Note immediately that a 1-body problem is not a special case of the general N-body problem, which has been formulated only for $N \geq 2$.

Finally, observe that each of the models to be studied is nonlinear. Linear models often have only limited life spans which end when refinement becomes essential.

Chapter 1

The 1-Body Problem

1.1. Nonlinear Oscillation

An important class of 1-body problems is found in the study of nonlinear oscillators. An oscillator is a body which moves up and back over all or part of a finite path. The prototype nonlinear oscillator is the swinging pendulum and in this section we turn attention to it. Other nonlinear oscillators can be treated in the fashion to be developed.

Consider a pendulum, as shown in Figure 1.1, which has mass m centered at P and is hinged at O. Assume that P is constrained to move on a circle of radius l whose center is O. Let θ be the angular measure, in radians, of the pendulum's deviation from the vertical. The problem is that of describing the motion of P after release from a position of rest.

It is known from laboratory experiments that the motion of the pendulum is damped and that the length of time between successive swings decreases.

Using cgs units, we reason analytically as follows. Assume that the motion of P is determined by Newton's dynamical equation

$$F = ma. \tag{1.1}$$

Circular arc NP has length $l\theta$, so that $a = \frac{d^2}{dt^2}(l\theta) = l\ddot{\theta}$. Thus, (1.1) becomes

$$F = ml\ddot{\theta}. \tag{1.2}$$

In considering the force F which acts on P, let F_1 be the gravitational component, so that

$$F_1 = -mg\sin\theta, \quad g > 0, \tag{1.3}$$

1

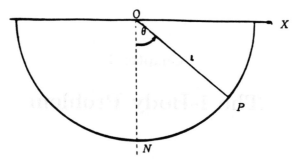

Figure 1.1.   A pendulum.

and let F_2 be a damping component of the form

$$F_2 = -c\dot{\theta}, \ c \text{ a nonnegative constant.} \tag{1.4}$$

Assume that these are the only forces whose effects are significant. Then

$$F = -mg\sin\theta - c\dot{\theta},$$

so that (1.2) reduces readily to

$$\ddot{\theta} + \frac{c}{ml}\dot{\theta} + \frac{g}{l}\sin\theta = 0 \tag{1.5}$$

The problem, then, is one of solving (1.5) subject to given initial conditions

$$\theta(0) = \alpha, \quad \dot{\theta}(0) = 0. \tag{1.6}$$

For illustrative purposes, let us consider the strongly damped pendulum motion defined by

$$\ddot{\theta} + (0.3)\dot{\theta} + \sin\theta = 0 \tag{1.7}$$

$$\theta(0) = \frac{1}{4}\pi, \quad \dot{\theta}(0) = 0. \tag{1.8}$$

No analytical method is known for constructing the exact solution of this problem. Numerically, then, set $t = x$ and $\theta = y$ and solve (1.7) with $\Delta t = 0.01$ using Kutta's fourth order formulas, which are given in generic form in Appendix I. The computation is carried out for 15000 time steps, that is, for 150 seconds of pendulum motion. The first 15.0 seconds of pendulum oscillation is shown in Figure 1.2, where the peak, or extreme, values 0.78540, -0.47647, 0.29335, -0.18156, 0.11259 occur at the times 0.00, 3.28, 6.49, 9.68, 12.86, respectively. The time required for the pendulum to travel

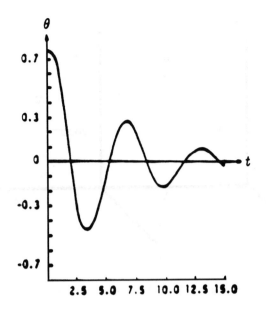

Figure 1.2.   Damped pendulum motion.

from one peak to another decreases monotonically and damping is present during the entire simulation, in agreement with experimentation.

Any attempt to linearize (1.7) results in an analytical solution which either does not damp out, or has a constant time interval between successive swings, or both.

1.2.   Concepts from Special Relativity

Interestingly enough, there exist some very important 1-body problems in Special Relativity. Let us then turn to a basic dynamical problem in Special Relativity and begin by discussing the very few concepts which will be required for the development. Incidentally, Special Relativity does not allow N-body problems for $N > 1$ because simultaneity is not a property of this branch of physics.

In Special Relativity one takes into account the time required for light to travel from a phenomenon being observed to the eye of the observer. Consider, then, two reference frames: a *lab* frame with Euclidean coordinates X, Y, Z and a *rocket* frame with Euclidean coordinates X', Y', Z', which coincide initially. In the frames one positions observers O and O' at

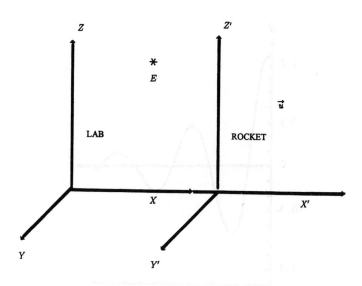

Figure 1.3.   Lab and Rocket frames.

their respective origins. At some initial time the observers have synchronized clocks. Assume the rocket frame is in motion in the X direction with speed u relative to the lab frame. (See Figure 1.3) Assume that $|u|$ is less than the speed of light.

An event E, like an exploding star, is observed by both O and O'. O records E as happening at (x, y, z) at time t, while O' records E as happening at (x', y', z') at time t'. Taking into account the time for light to travel to the eyes of the observers, these variables are related by the Lorentz transformation (Bergmann (1976)):

$$x' = \frac{c(x - ut)}{(c^2 - u^2)^{1/2}}, \quad y' = y, \quad z' = z,$$

$$t' = \frac{(c^2 t - ux)}{c(c^2 - u^2)^{1/2}}, \quad |u| < c, \tag{1.9}$$

or, equivalently,

$$x = \frac{c(x' + ut')}{(c^2 - u^2)^{1/2}}, \quad y = y', \quad z = z',$$

$$t = \frac{(c^2 t' + ux')}{c(c^2 - u^2)^{1/2}}, \quad |u| < c, \tag{1.10}$$

in which c is the speed of light.

For *covariance* relative to the Lorentz transformation, Einstein showed that for the motion of a particle P of rest mass m_0,

$$F = \frac{d}{dt}(mv), \quad m = \frac{cm_0}{(c^2 - v^2)^{1/2}}, \quad |v| < c, \qquad \text{(lab)} \qquad (1.11)$$

maps under the Lorentz transformation into

$$F' = \frac{d}{dt'}(m'v'), \quad m' = \frac{cm_0}{(c^2 - v'^2)^{1/2}}, \quad |v'| < c, \qquad \text{(rocket)} \qquad (1.12)$$

that is, the laws of motion are the *same* in both the lab frame and the rocket frame. For future use and because of its basic importance, let us actually prove this result.

We first define the continuum concepts of velocity and acceleration. In the lab frame, set

$$v = \frac{dx}{dt}, \quad a = \frac{dv}{dt}, \quad |v| < c, \qquad (1.13)$$

while in the rocket frame set

$$v' = \frac{dx'}{dt'}, \quad a' = \frac{dv'}{dt'}, \quad |v'| < c. \qquad (1.14)$$

To relate v and v', we have from (1.9)

$$v' = \frac{dx'}{dt'} = \frac{c^2(dx - udt)}{c^2 dt - udx} \qquad (1.15)$$

so that

$$v' = \frac{c^2(v - u)}{c^2 - uv}. \qquad (1.16)$$

Equivalently, from (1.10), one finds

$$v = \frac{c^2(v' + u)}{c^2 + uv'}. \qquad (1.17)$$

Similarly, the relationship between a and a' is found to be

$$a' = \frac{c^3(c^2 - u^2)^{3/2}}{(c^2 - uv)^3} a, \qquad (1.18)$$

or, equivalently,

$$a = \frac{c^3(c^2 - u^2)^{3/2}}{(c^2 + uv')^3} a'. \qquad (1.19)$$

We are now ready to prove the Einstein result which is formulated in the following theorem.

Theorem 1.1.   *Let a particle P be in motion along the X axis in the lab and along the X' axis in the rocket. In the lab frame let the mass m of P be given by*

$$m = \frac{cm_0}{(c^2 - v^2)^{1/2}}, \quad |v| < c, \tag{1.20}$$

where m_0 is a positive constant called the rest mass of P and v is the speed of P in the lab. In the rocket frame let the mass m' of P be given by

$$m' = \frac{cm_0}{(c^2 - (v')^2)^{1/2}}, \quad |v'| < c, \tag{1.21}$$

where m_0 is the same constant as in (1.20) and v' is the speed of P in the rocket. Let a force F be applied to P in the lab. In rocket coordinates denote the force by F', so that

$$F = F'.$$

Then, if in the lab the equation of motion is given by

$$F = \frac{d}{dt}(mv), \tag{1.22}$$

it follows that in the rocket the equation of motion of P is

$$F' = \frac{d}{dt'}(m'v'). \tag{1.23}$$

Proof.   From (1.22) and (1.20)

$$F = v\frac{dm}{dt} + m\frac{dv}{dt}$$

$$= \frac{v^2\,ma}{c^2 - v^2} + ma$$

so that

$$F = \left(\frac{c^2}{c^2 - v^2}\right)ma. \tag{1.24}$$

From (1.15) and (1.13), then, we must have

$$F' = \left(\frac{c^2}{c^2 - (v')^2}\right)m'a'. \tag{1.25}$$

Since $F = F'$, the proof will follow if

$$\left(\frac{c^2}{c^2 - v^2}\right) ma \equiv \left(\frac{c^2}{c^2 - (v')^2}\right) m'a'. \tag{1.26}$$

However, substitution of (1.16), (1.18) and (1.21) into the right side of (1.26) yields, quite remarkably, that the identity is valid and the theorem is proved. $\quad\square$

1.3. Relativistic Oscillation

Let us consider now a particle P which oscillates on the X axis in the lab frame. Assume that the force F on P is one whose magnitude depends only on the x coordinate of P. Then, let

$$F = f(x). \tag{1.27}$$

Assume that initially, that is, at time $t = 0, P$ is at x_0 and has speed v_0. Then the equation of motion for P in the lab frame is

$$\frac{d}{dt}(mv) = f(x), \tag{1.28}$$

or,

$$\left(\frac{c^2}{c^2 - v^2}\right) ma = f(x), \tag{1.29}$$

or,

$$c^2 m\ddot{x} = f(x)(c^2 - \dot{x}^2). \tag{1.30}$$

In turn, the latter equation reduces to

$$c^3 m_0 \ddot{x} = f(x)(c^2 - \dot{x}^2)^{3/2},$$

so that, finally, we find

$$\ddot{x} - \frac{f(x)}{c^3 m_0}(c^2 - \dot{x}^2)^{3/2} = 0, \tag{1.31}$$

is the differential equation one has to solve in the lab frame, given the initial conditions

$$x(0) = x_0, \quad \dot{x}(0) = v_0. \tag{1.32}$$

In general, (1.31) cannot be solved in closed form, so that the observer in the lab must now introduce a computer to approximate the solution. However, the observer in the rocket frame also observes the motion of P, but in his coordinate system. His equation and initial conditions are found by applying (1.10), (1.17) and (1.19) to (1.31) and (1.32). Thus, he too will be confronted with a problem which requires a computer and so a computer identical to that in the lab is now introduced into the rocket.

The fundamental problem which now arises is: How should the computations be done in the lab and rocket so that the physics of Special Relativity is preserved, that is, so that the *numerical* results will be related by the Lorentz transformation. We show now how this can be done.

1.4.   Numerical Methodology

For $\Delta t > 0$, let $t_k = k\Delta t, k = 0, 1, 2, 3 \ldots$. Let t'_k correspond to t_k by the Lorentz transformation. At t_k let P be at (x_k, y_k, z_k) in the lab and at (x'_k, y'_k, z'_k) in the rocket. These are also related by the Lorentz transformation. Define

$$v_k = \frac{\Delta x_k}{\Delta t_k} = \frac{x_{k+1} - x_k}{t_{k+1} - t_k}, \quad a_k = \frac{\Delta v_k}{\Delta t_k} = \frac{v_{k+1} - v_k}{t_{k+1} - t_k}, \quad \text{(LAB)} \qquad (1.33)$$

$$v'_k = \frac{\Delta x'_k}{\Delta t'_k} = \frac{x'_{k+1} - x'_k}{t'_{k+1} - t'_k}, \quad a'_k = \frac{\Delta v'_k}{\Delta t'_k} = \frac{v'_{k+1} - v'_k}{t'_{k+1} - t'_k} \quad \text{(ROCKET)} \quad (1.34)$$

Then corresponding to (1.16)–(1.19), one has by direct substitution that

$$v'_k = \frac{c^2(v_k - u)}{c^2 - uv_k}, \quad v_k = \frac{c^2(v'_k + u)}{c^2 + uv'_k} \qquad (1.35)$$

$$a'_k = \frac{c^3(c^2 - u^2)^{3/2}}{(c^2 - uv_k)^2(c^2 - uv_{k+1})} a_k,$$

$$a_k = \frac{c^3(c^2 - u^2)^{3/2}}{(c^2 + uv'_k)^2(c^2 + uv'_{k+1})} a'_k. \qquad (1.36)$$

In the limit (1.33)–(1.36) converge to (1.13), (1.14) and (1.16)–(1.19). Our problem now is to choose an approximation for

$$F = \frac{d}{dt}(mv) \qquad (1.37)$$

in the lab which will transform covariantly into the rocket. The clue for such a choice comes from (1.24) which is equivalent to (1.22). What we choose at t_k is the approximation

$$F_k = \frac{c^2 m_k}{[(c^2 - v_k^2)(c^2 - v_{k+1}^2)]^{1/2}} \frac{\Delta v_k}{\Delta t_k}, \quad m_k = \frac{cm_0}{(c^2 - v_k^2)^{1/2}}. \quad (1.38)$$

Note first that in the limit, (1.38) converges to (1.24). What we must prove is that if $F_k = F_k'$ and if in the rocket

$$m_k' = \frac{cm_0}{(c^2 - v_k'^2)^{1/2}} \quad (1.39)$$

then in the rocket

$$F_k' = \frac{c^2 m_k'}{[(c^2 - v_k'^2)(c^2 - v_{k+1}'^2)]^{1/2}} \frac{\Delta v_k'}{\Delta t_k'}. \quad (1.40)$$

Theorem 1.2. (*Covariance.*) *Under the Lorentz transformation*

$$F_k = \frac{c^2 m_k}{[(c^2 - v_k^2)(c^2 - v_{k+1}^2)]^{1/2}} \frac{\Delta v_k}{\Delta t_k}, \quad m_k = \frac{cm_0}{(c^2 - v_k^2)^{1/2}} \quad (1.41)$$

transforms into

$$F_k' = \frac{c^2 m_k'}{[(c^2 - v_k'^2)(c^2 - v_{k+1}'^2)]^{1/2}} \frac{\Delta v_k'}{\Delta t_k'}, \quad m_k' = \frac{cm_0}{(c^2 - v_k'^2)^{1/2}}. \quad (1.42)$$

Moreover, (1.41) and (1.42) converge to the Einstein equations as the time steps converge to zero, that is, consistency is valid.

Proof. The proof is completely analogous to that of Theorem 1.1, but uses (1.33)–(1.36) instead of (1.16)–(1.19).

Interestingly enough, (1.33), (1.34), (1.41), (1.42) can be solved explicitly to yield

$$x_{k+1} = x_k + (\Delta t_k) v_k \quad (1.43)$$

$$v_{k+1} = \frac{v_k + (\Delta t_k)(1 - (v_k)^2) F_k \sqrt{1 - (v_k)^2 + (\Delta t_k)^2 (1 - (v_k)^2)^2 (F_k)^2}}{1 + (\Delta t_k)^2 (1 - (v_k)^2)^2 (F_k)^2} \quad (1.44)$$

$$x_{k+1}' = x_k' + (\Delta t_k') v_k' \quad (1.45)$$

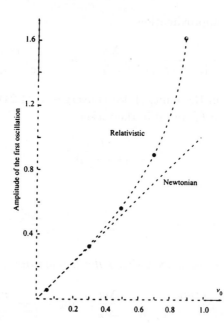

Figure 1.4.   Comparative harmonic motion.

$$v'_{k+1} = \frac{v'_k + (\Delta t'_k)(1 - (v'_k)^2)F'_k \sqrt{1 - (v'_k)^2 + (\Delta t'_k)^2(1 - (v'_k)^2)^2(F'_k)^2}}{1 + (\Delta t'_k)^2(1 - (v'_k)^2)^2(F'_k)^2}.$$

$$(1.46)$$

from which computations are done readily.   □

For example, if we define a relativistic harmonic oscillator as a particle for which $F(x) = -K^2x$, where K is a nonzero constant, then we can apply (1.43), (1.44) in the lab to study the resulting dynamical behavior of the particle from given initial data. Thus, if we normalize by setting $K = m_0 = c = 1$, and assume that $x(0) = x_0 = 0$. then Figure 1.4 shows how the amplitude of the relativistic harmonic oscillator deviates from that of the Newtonian harmonic oscillator with increasing v_0.

Finally we have a major theorem.

Theorem 1.3. *For simplicity only, assume the normalization $m_0 = c = 1$. Then, using (1.43) and (1.44) numerically in the lab and using (1.45) and (1.46) numerically in the rocket results in numerical results which are related by the Lorentz transformation.*

Proof. In general, for an arbitrary force $f(x)$,

$$F_k = f(x_k) = f\left(\frac{x'_k + ut'_k}{(1-u^2)^{1/2}}\right) = F'_k. \tag{1.47}$$

We wish to prove that

$$x'_{k+1} = \frac{x_{k+1} - ut_{k+1}}{(1-u^2)^{1/2}} \tag{1.48}$$

$$v'_{k+1} = \frac{v_{k+1} - u}{1 - uv_{k+1}}. \tag{1.49}$$

Assume that x_0, x'_0, v_0, v'_0, the given initial data, are related by the Lorentz transformation. We will then prove (1.48), (1.49) for $k = 0$. This will be sufficient, by induction, to establish (1.48), (1.49). Thus,

$$x'_0 = \frac{x_0 - ut_0}{(1-u^2)^{1/2}}, \quad v'_0 = \frac{v_0 - u}{1 - uv_0}. \tag{1.50}$$

Now,

$$x'_1 = x'_0 + v'_0(\Delta t'_0). \tag{1.51}$$

Let x_1^* correspond to x'_1 by the Lorentz transformation, so that

$$x'_1 = \frac{x_1^* - ut_1}{(1-u^2)^{1/2}}.$$

We want to show that $x_1^* = x_1$.

We know that

$$t'_0 = \frac{t_0 - ux_0}{(1-u^2)^{1/2}}, \quad t'_1 = \frac{t_1 - ux_1^*}{(1-u^2)^{1/2}}. \tag{1.52}$$

Thus,

$$\frac{x_1^* - ut_1}{(1-u^2)^{1/2}} = x'_0 + v'_0(\Delta t'_0)$$

$$= \frac{x_0 - ut_0}{(1-u^2)^{1/2}} + \left(\frac{v_0 - u}{1 - uv_0}\right)\left(\frac{t_1 - ux_1^*}{(1-u^2)^{1/2}} - \frac{t_0 - ux_0}{(1-u^2)^{1/2}}\right).$$

Since $(1 - u^2) > 0$, it then follows that

$$x_1^* - ut_1 = x_0 - ut_0 + \frac{v_0 - u}{1 - uv_0}(t_1 - t_0 + ux_0 - ux_1^*)$$

or

$$(x_1^* - ut_1)(1 - uv_0) = (x_0 - ut_0)(1 - uv_0) + (v_0 - u)(t_1 - t_0) + (v_0 - u)(ux_0 - ux_1^*),$$

which simplifies to

$$(x_1^* - x_0)(1 - u^2) = v_0(1 - u^2)(t_1 - t_0).$$

Thus

$$x_1^* - x_0 = v_0(t_1 - t_0)$$

so that

$$x_1^* = x_0 + v_0(t_1 - t_0).$$

Thus, $x_1^* = x_1$.

Next, for $k = 0$,

$$v_1' = \frac{v_0' + (\Delta t_0')(1 - (v_0')^2)F_0'\sqrt{1 - (v_0')^2 + (\Delta t_0')^2(1 - (v_0')^2)^2(F_0')^2}}{1 + (\Delta t_0')^2(1 - (v_0')^2)^2(F_0')^2}.$$

$$(1.53)$$

From (1.40), (1.50) and (1.51), substitution of

$$t_0' = \frac{t_0 - ux_0}{(1 - u^2)^{1/2}}, \quad t_1' = \frac{t_1 - ux_1}{(1 - u^2)^{1/2}}$$

$$F_0' = F_0 = \frac{v_1 - v_0}{(t_1 - t_0)(1 - v_0^2)(1 - v_1^2)^{1/2}}$$

$$x_0' = \frac{x_0 - ut_0}{(1 - u^2)^{1/2}}, \quad v_0' = \frac{v_0 - u}{1 - uv_0}$$

into (1.53) yields

$$v_1' = \frac{(v_1 - u)[(1 - v_0v_1)(1 - uv_0) + (v_1 - v_0)(u - v_0)]}{(1 - uv_1)[(1 - v_0v_1)(1 - uv_0) + (v_1 - v_0)(u - v_0)]}.$$

Finally, we will have the desired result

$$v_1' = \frac{(v_1 - u)}{(1 - uv_1)}$$

provided the terms in the brackets, which are identical, are not zero, and we will show this next.

Note first that since we have normalized so that $|u| < 1$ and $|v| < 1$, then

$$|u - v| < 1 - uv.$$

However,

$$(1 - v_0v_1)(1 - uv_0) + (v_1 - v_0)(u - v_0) > 0$$

since

$$|v_1 - v_0| < 1 - v_0 v_1$$
$$|u - v_0| < 1 - uv_0,$$

so that

$$-(v_1 - v_0)(u - u_0) \leq |v_1 - v_0||u - u_0| < (1 - v_0 v_1)(1 - uv_0).$$

Hence,

$$(1 - v_0 v_1)(1 - uv_0) + (v_1 - v_0)(u - v_0) > 0.$$

$\square$

1.5.   Remark

There exist 1-body problems which can be solved in closed form. These include, for example, the motion of a planet around a fixed sun and certain problems in ballistics theory (Synge and Griffith (1942)). However, these closed form solutions require severe assumptions which are usually unrealistic. Inclusion of one or more important relevant constraints requires numerical methodology like that used in Section 1.1.

Chapter 2

N-Body Problems with $2 \leq N \leq 100$

2.1. Introduction

In general, we will wish to consider N-body problems in which N may be relatively large and relatively small. N-body problems with $2 \leq N \leq 100$ will be considered to be *small* and we will begin with these. For $N = 2$, and under important restrictions, it may be possible to solve related problems in closed form. This is the case, for example, in astromechanics (van de Kamp (1964)). Inclusion of various important constraints, however, then demands numerical methodology.

Now, if N is small, we would like to do a very good job in solving the N-body problem. By this we mean that we would like not only to solve the problem with accuracy, but we would also like to preserve numerically any basic physical invariants of the system. To do this in detail, we concentrate theoretically and computationally on the 3-body problem, because it contains *all* the difficulties of the general N-body problem. The entire discussion extends in a natural way to the general N-body problem, and, in particular, to the more simplistic 2-body problem.

For $i = 1, 2, 3$, let P_i of mass m_i be at $\vec{r}_i = (x_i, y_i, z_i)$ at time t. Let the positive distance between P_i and P_j, $i \neq j$, be $r_{ij} = r_{ji}$. Let $\phi = \phi_{ij} = \phi(r_{ij})$, given in ergs, be a potential for the pair P_i, P_j. Then the force on P_i due to P_j is

$$\vec{F}_i = -\frac{\partial \phi}{\partial r_{ij}} \frac{\vec{r}_i - \vec{r}_j}{r_{ij}},$$

and in this section we assume Newtonian dynamical differential equations for the 3-body problem, and these are

$$m_i \frac{d^2 \vec{r}_i}{dt^2} = -\frac{\partial \phi}{\partial r_{ij}} \frac{\vec{r}_i - \vec{r}_j}{r_{ij}} - \frac{\partial \phi}{\partial r_{ik}} \frac{\vec{r}_i - \vec{r}_k}{r_{ik}}, \quad i = 1, 2, 3 \qquad (2.1)$$

where $i = 1$ implies $j = 2, k = 3$; $i = 2$ implies $j = 1, k = 3$; $i = 3$ implies $j = 1, k = 2$.

The following summary theorem incorporates several well known results.

Theorem 2.1. (Goldstein (1980)) *System* (2.1) *conserves energy, linear momentum, and angular momentum. It is also covariant under translation, rotation, and uniform relative motion of coordinate frames.*

2.2. Numerical Methodology

In general, (2.1) will be nonlinear and will require numerical methodology. In order to solve an initial value problem for (2.1) numerically, we first rewrite it as the equivalent first order system

$$\frac{d\vec{r}_i}{dt} = \vec{v}_i, \quad i = 1, 2, 3 \tag{2.2}$$

$$m_i \frac{d\vec{v}_i}{dt} = -\frac{\partial \phi}{\partial r_{ij}} \frac{\vec{r}_i - \vec{r}_j}{r_{ij}} - \frac{\partial \phi}{\partial r_{ik}} \frac{\vec{r}_i - \vec{r}_k}{r_{ik}}, \quad i = 1, 2, 3. \tag{2.3}$$

Our numerical formulation now proceeds as follows. For $\Delta t > 0$, let $t_n = n(\Delta t), n = 0, 1, 2, \ldots$. At time t_n, let P_i be at $\vec{r}_{i,n} = (x_{i,n}, y_{i,n}, z_{i,n})$ with velocity $\vec{v}_{i,n} = (v_{i,x,n}, v_{i,y,n}, v_{i,z,n})$. Denote the distances $\|P_1 P_2\|$, $\|P_1 P_3\|$, $\|P_2 P_3\|$ by $r_{12,n}$, $r_{13,n}$, $r_{23,n}$, respectively. Differential equations (2.2) and (2.3) are now approximated, respectively, by the difference equations

$$\frac{\vec{r}_{i,n+1} - \vec{r}_{i,n}}{\Delta t} = \frac{\vec{v}_{i,n+1} + \vec{v}_{i,n}}{2} \tag{2.4}$$

$$m_i \frac{\vec{v}_{i,n+1} - \vec{v}_{i,n}}{\Delta t} = -\frac{\phi(r_{ij,n+1}) - \phi(r_{ij,n})}{r_{ij,n+1} - r_{ij,n}} \frac{\vec{r}_{i,n+1} + \vec{r}_{i,n} - \vec{r}_{j,n+1} - \vec{r}_{j,n}}{r_{ij,n+1} + r_{ij,n}}$$
$$- \frac{\phi(r_{ik,n+1}) - \phi(r_{ik,n})}{r_{ik,n+1} - r_{ik,n}} \frac{\vec{r}_{i,n+1} + \vec{r}_{i,n} - \vec{r}_{k,n+1} - \vec{r}_{k,n}}{r_{ik,n+1} + r_{ik,n}}. \tag{2.5}$$

Note that the force is approximated, not the potential. We take the very same potential as in continuum mechanics, the significance of which will be seen shortly. Consistency follows immediately as $\Delta t \to 0$. Also note that, for the present, we assume in (2.5) that $r_{lm,n+1} \neq r_{lm,n}$, for any choices of l, m.

System (2.4), (2.5) consists of 18 implicit equations for the unknowns $x_{i,n+1}$, $y_{i,n+1}$, $z_{i,n+1}$, $v_{i,x,n+1}$, $v_{i,y,n+1}$, $v_{i,z,n+1}$ in the 18 knowns $x_{i,n}$, $y_{i,n}$, $z_{i,n}$, $v_{i,x,n}$, $v_{i,y,n}$, $v_{i,z,n}$ and is solvable readily by Newton's method, as described in Appendix II.

2.3. Conservation Laws

Because of its physical significance, let us show now that the numerical solution generated by (2.4) and (2.5) conserves the same energy, linear momentum, and angular momentum as does (2.1).

Consider first energy conservation. Define

$$W_N = \sum_{n=0}^{N-1} \left\{ \sum_{i=1}^{3} m_i(\vec{r}_{i,n+1} - \vec{r}_{i,n}) \cdot (\vec{v}_{i,n+1} - \vec{v}_{i,n})/\Delta t \right\}. \qquad (2.6)$$

Note immediately relative to (2.6) that, since we are considering specifically the three-body problem, the symbol N in summation (2.6) is, in this section only, a numerical time index. Then insertion of (2.4) into (2.6) and simplification yields

$$W_N = \frac{1}{2}m_1(v_{1,N})^2 + \frac{1}{2}m_2(v_{2,N})^2 + \frac{1}{2}m_3(v_{3,N})^2$$
$$- \frac{1}{2}m_1(v_{1,0})^2 - \frac{1}{2}m_2(v_{2,0})^2 - \frac{1}{2}m_3(v_{3,0})^2,$$

so that

$$W_N = K_N - K_0. \qquad (2.7)$$

Insertion of (2.5) into (2.6) implies, with some tedious algebraic manipulation,

$$W_N = \sum_{n=0}^{N-1} (-\phi_{12,n+1} - \phi_{13,n+1} - \phi_{23,n+1} + \phi_{12,n} + \phi_{13,n} + \phi_{23,n})$$

so that

$$W_N = -\phi_N + \phi_0. \qquad (2.8)$$

Elimination of W_N between (2.7) and (2.8) then yields conservation of energy, that is,

$$K_N + \phi_N = K_0 + \phi_0, \quad N = 1, 2, 3, \ldots.$$

Moreover, since K_0 and ϕ_0 depend only on initial data, it follows that K_0 and ϕ_0 are the same in both the continuous and the discrete cases, so that the energy conserved by the numerical method is exactly that of the continuous system. Note, in addition, that the proof is independent of Δt. Thus, we have proved the following theorem.

Theorem 2.2.   *Independently of* Δt, *the numerical method of Section 2.2 is energy conserving, that is,*

$$K_N + \phi_N = K_0 + \phi_0, \quad N = 1, 2, 3, \ldots.$$

To show the conservation of linear momentum, we proceed as follows. The linear momentum $\vec{M}_i(t_n) = \vec{M}_{i,n}$ of P_i at t_n is defined to be the vector

$$\vec{M}_{i,n} = m_i(v_{i,n,x}, v_{i,n,y}, v_{i,n,z}). \tag{2.9}$$

The linear momentum $\vec{M}_n$ of the three-body system at time t_n is defined to be the vector

$$\vec{M}_n = \sum_{i=1}^{3} \vec{M}_{i,n}. \tag{2.10}$$

Now, from (2.5),

$$m_1(\vec{v}_{1,n+1} - \vec{v}_{1,n}) + m_2(\vec{v}_{2,n+1} - \vec{v}_{2,n}) + m_3(\vec{v}_{3,n+1} - \vec{v}_{3,n}) \equiv \vec{0}.$$

Thus, for $n = 0, 1, 2, \ldots,$

$$m_1(v_{1,n+1,x} - v_{1,n,x}) + m_2(v_{2,n+1,x} - v_{2,n,x}) + m_3(v_{3,n+1,x} - v_{3,n,x}) = 0. \tag{2.11}$$

Summing both sides of (2.11) from $n = 0$ to $n = N - 1$ implies

$$m_1 v_{1,N,x} + m_2 v_{2,N,x} + m_3 v_{3,N,x} = C_1, \quad N \geq 1 \tag{2.12}$$

in which

$$m_1 v_{1,0,x} + m_2 v_{2,0,x} + m_3 v_{3,0,x} = C_1. \tag{2.13}$$

Similarly,

$$m_1 v_{1,N,y} + m_2 v_{2,N,y} + m_3 v_{3,N,y} = C_2 \tag{2.14}$$

$$m_1 v_{1,N,z} + m_2 v_{2,N,z} + m_3 v_{3,N,z} = C_3 \tag{2.15}$$

in which

$$m_1 v_{1,0,y} + m_2 v_{2,0,y} + m_3 v_{3,0,y} = C_2 \tag{2.16}$$

$$m_1 v_{1,0,z} + m_2 v_{2,0,z} + m_3 v_{3,0,z} = C_3. \tag{2.17}$$

Thus,

$$\vec{M}_n = \sum_{i=1}^{3} \vec{M}_{i,n} = (C_1, C_2, C_3) = \vec{M}_0, \quad n = 1, 2, 3, \ldots,$$

which is the classical law of conservation of linear momentum. Note that $\vec{M}_0$ depends only on the initial data. Thus we have the following theorem.

Theorem 2.3. *Independently of Δt, the numerical method of Section 2.2 conserves linear momentum, that is,*

$$\vec{M}_n = \vec{M}_0, \quad n = 1, 2, 3, \ldots.$$

To show the conservation of angular momentum, we proceed as follows. The angular momentum $\vec{L}_{i,n}$ of P_i at t_n is defined to be the cross product vector

$$\vec{L}_{i,n} = m_i(\vec{r}_{i,n} \times \vec{v}_{i,n}). \tag{2.18}$$

The angular momentum of a three-body system at t_n is defined to be the vector

$$\vec{L}_n = \sum_{i=1}^{3} \vec{L}_{i,n}. \tag{2.19}$$

It then follows readily that

$$\vec{L}_{i,n+1} - \vec{L}_{i,n}$$
$$= m_i(\vec{r}_{i,n+1} + \vec{r}_{i,n}) \times (\vec{v}_{i,n+1} - \vec{v}_{i,n})$$
$$= m_i \left[(\vec{r}_{i,n+1} - \vec{r}_{i,n}) \times \frac{1}{2}(\vec{v}_{i,n+1} + \vec{v}_{i,n}) \right.$$
$$\left. + \frac{1}{2}(\vec{r}_{i,n+1} + \vec{r}_{i,n}) \times (\vec{v}_{i,n+1} - \vec{v}_{i,n}) \right]$$
$$= m_i \left[(\vec{r}_{i,n+1} - \vec{r}_{i,n}) \times \frac{1}{\Delta t}(\vec{r}_{i,n+1} - \vec{r}_{i,n}) + \frac{1}{2}(\vec{r}_{i,n+1} + \vec{r}_{i,n}) \times \vec{a}_{i,n}\Delta t \right]$$
$$= \frac{1}{2}(\Delta t)(\vec{r}_{i,n+1} + \vec{r}_{i,n}) \times \vec{F}_{i,n}.$$

For notational simplicity, set

$$\vec{T}_{i,n} = \frac{1}{2}(\vec{r}_{i,n+1} + \vec{r}_{i,n}) \times \vec{F}_{i,n}.$$

It follows readily, with some algebraic manipulation, that

$$\vec{T}_n = \vec{T}_{1,n} + \vec{T}_{2,n} + \vec{T}_{3,n} = 0.$$

Thus, one finds

$$\vec{L}_{n+1} - \vec{L}_n = \vec{0}, \quad n = 0, 1, 2, 3, \ldots,$$

so that

$$\vec{L}_n = \vec{L}_0, \quad n = 1, 2, 3, \ldots,$$

which implies, independently of Δt, the conservation of angular momentum. Note again that $\vec{L}_0$ depends only on the initial data. Thus the following theorem has been proved.

Theorem 2.4. *Independently of Δt, the numerical method of Section 2.2 conserves angular momentum, that is*

$$\vec{L}_n = \vec{L}_0, \quad n = 1, 2, 3, \ldots.$$

2.4. Covariance

We begin the discussion of Newtonian covariance by stating the basic concepts. When a dynamical equation is structurally invariant under a transformation, the equation is said to be *covariant or symmetric*. The transformations we will consider are the basic ones, namely, translation, rotation, and uniform relative motion. We will concentrate on two dimensional systems, because the related techniques and results extend directly to three dimensions. A general Newtonian force will be considered. Finally, we will concentrate on the motion of a single particle P of mass m, with extension to the N-body problem following in a natural way. And though the assumptions just made may seem to be excessive, it will be seen shortly that they render the required mathematical methodology readily transparent.

Suppose now that a particle P of mass m is in motion in the XY plane and that for $\Delta t > 0$ its motion from given initial data is determined by a force $\vec{F}(t_n) = \vec{F}_n = (F_{n,x}, F_{n,y})$ and by the dynamical difference equations

$$F_{n,x} = m(v_{n+1,x} - v_{n,x})/(\Delta t) \tag{2.20}$$

$$F_{n,y} = m(v_{n+1,y} - v_{n,y})/(\Delta t). \tag{2.21}$$

The fundamental problem that we now consider is as follows. Let $x = f_1(x^*, y^*), y = f_2(x^*, y^*)$ be a change of coordinates. Under this transformation, let $F_{n,x} = F^*_{n,x^*}, F_{n,y} = F^*_{n,y^*}$. Then we will want to prove that in the X^*Y^* system the dynamical equations of motion are

$$F_{n,x^*}^* = m(v_{n+1,x^*} - v_{n,x^*})/(\Delta t) \tag{2.22}$$

$$F_{n,y^*}^* = m(v_{n+1,y^*} - v_{n,y^*})/(\Delta t), \tag{2.23}$$

which will establish covariance.

In consistency with (2.4), we assume that

$$\frac{x_{n+1} - x_n}{\Delta t} = \frac{v_{n+1,x} + v_{n,x}}{2}, \quad \frac{x_{n+1}^* - x_n^*}{\Delta t} = \frac{v_{n+1,x^*} + v_{n,x^*}}{2} \tag{2.24}$$

$$\frac{y_{n+1} - y_n}{\Delta t} = \frac{v_{n+1,y} + v_{n,y}}{2}, \quad \frac{y_{n+1}^* - y_n^*}{\Delta t} = \frac{v_{n+1,y^*} + v_{n,y^*}}{2}. \tag{2.25}$$

Relative to (2.24) and (2.25), the following lemma will be of value.

Lemma 2.1. *Equations (2.24) and (2.25) imply*

$$v_{1,x} = \frac{2}{\Delta t}(x_1 - x_0) - v_{0,x}; \quad v_{1,x^*} = \frac{2}{\Delta t}(x_1^* - x_0^*) - v_{0,x^*} \tag{2.26}$$

$$v_{1,y} = \frac{2}{\Delta t}(y_1 - y_0) - v_{0,y}; \quad v_{1,y^*} = \frac{2}{\Delta t}(y_1^* - y_0^*) - v_{0,y^*} \tag{2.27}$$

$$v_{n,x} = \frac{2}{\Delta t}\left[x_n + (-1)^n x_0 + 2\sum_{j=1}^{n-1}(-1)^j x_{n-j} \right] + (-1)^n v_{0,x}, \quad n \geq 2 \tag{2.28}$$

$$v_{n,x^*} = \frac{2}{\Delta t}\left[x_n^* + (-1)^n x_0^* + 2\sum_{j=1}^{n-1}(-1)^j x_{n-j}^* \right] + (-1)^n v_{0,x^*}, \quad n \geq 2 \tag{2.29}$$

$$v_{n,y} = \frac{2}{\Delta t}\left[y_n + (-1)^n y_0 + 2\sum_{j=1}^{n-1}(-1)^j y_{n-j} \right] + (-1)^n v_{0,y}, \quad n \geq 2 \tag{2.30}$$

$$v_{n,y^*} = \frac{2}{\Delta t}\left[y_n^* + (-1)^n y_0^* + 2\sum_{j=1}^{n-1}(-1)^j y_{n-j}^* \right] + (-1)^n v_{0,y^*}, \quad n \geq 2. \tag{2.31}$$

Proof. Equations (2.26) follow directly from (2.24) with $n = 0$. Equations (2.27) follow directly from (2.25) with $n = 0$. Equations (2.28)–(2.31) follow readily by mathematical induction. $\qquad \square$

Theorem 2.5.   *Equations* (2.20) *and* (2.21) *are covariant relative to the translation*

$$x^* = x - a, \quad y^* = y - b; \quad a, b \text{ constants.}$$

Proof.   Define $v_{0,x} = v_{0,x^*}, v_{0,y} = v_{0,y^*}$. Then, from (2.26) in Lemma 2.1,

$$v_{1,x} = \frac{2}{\Delta t}\left[(x_1^* + a) - (x_0^* + a)\right] - v_{0,x^*} = v_{1,x^*}.$$

Similarly,

$$v_{1,y} = v_{1,y^*}$$

For $n > 1, (2.28)$ and (2.29) in Lemma 2.1 yield

$$v_{n,x} = \frac{2}{\Delta t}\left[(x_n^* + a) + (-1)^n(x_0^* + a) + 2\sum_{j=1}^{n-1}(-1)^j(x_{n-j}^* + a)\right]$$
$$+ (-1)^n v_{0,x^*}. \tag{2.32}$$

However, by the lemma, for n both odd and even, (2.32) implies

$$v_{n,x} = v_{n,x^*}.$$

Similarly,

$$v_{n,y} = v_{n,y^*}.$$

Thus, for all $n = 0, 1, 2, 3, \ldots$

$$v_{n,x} = v_{n,x^*},$$
$$v_{n,y} = v_{n,y^*}.$$

Thus,

$$F_{n,x^*}^* = F_{n,x} = m\frac{v_{n+1,x} - v_{n,x}}{\Delta t} = m\frac{v_{n+1,x^*} - v_{n,x^*}}{\Delta t}.$$

Similarly,

$$F_{n,y^*}^* = m\frac{v_{n+1,y^*} - v_{n,y^*}}{\Delta t},$$

and the theorem is proved.   □

Theorem 2.6.   *Under the rotation*

$$\begin{cases} x^* = x \cos \theta + y \sin \theta \\ y^* = y \cos \theta - x \sin \theta \end{cases} \tag{2.33}$$

where θ is the smallest positive angle measured counterclockwise from the X to the X^* axis, Eqs. (2.20), (2.21) are covariant.

Proof.   The proof follows along the same lines as that of Theorem 2.5 after one defines

$$\begin{cases} v_{0,x^*} = v_{0,x} \cos \theta + v_{0,y} \sin \theta \\ v_{0,y^*} = v_{0,y} \cos \theta - v_{0,x} \sin \theta \end{cases} \tag{2.34}$$

and notes that

$$\begin{cases} F_{n,x^*}^* = F_{n,x} \cos \theta + F_{n,y} \sin \theta \\ F_{n,y^*}^* = F_{n,y} \cos \theta - F_{n,x} \sin \theta \end{cases}. \tag{2.35}$$

$\square$

Theorem 2.7.   *Under relative uniform motion of coordinate systems, Eqs. (2.20), (2.21) are covariant.*

Proof.   Consider first motion in one dimension. Assume then that the X and X^* axes are in relative motion defined by

$$x_n^* = x_n - ct_n, \quad n = 0, 1, 2, 3, \ldots, \tag{2.36}$$

in which c is a positive constant. If $v_{0,x}$ is the initial velocity of P along the X axis, define v_{0,x^*} along the X^* axis by

$$v_{0,x^*} = v_{0,x} - c. \tag{2.37}$$

Hence, for $n = 1$,

$$v_{1,x} = \frac{2}{\Delta t} \left[(x_1^* + ct_1) - (x_0^* + ct_0) \right] - v_{0,x} = v_{1,x^*} + c. \tag{2.38}$$

For $n > 1$,

$$v_{n,x} = \frac{2}{\Delta t} \left\{ x_n^* + (-1)^n x_0^* + 2 \sum_{j=1}^{n-1} (-1)^j x_{n-j}^* \right\} + (-1)^n v_{0,x}$$

$$+ \frac{2c}{\Delta t} \left\{ t_n + (-1)^n t_0 + 2 \sum_{j=1}^{n-1} (-1)^j t_{n-j} \right\}.$$

But, it follows readily that

$$t_n + (-1)^n t_0 + 2 \sum_{j=1}^{n-1} (-1)^j t_{n-j} = \begin{cases} 0, & n \text{ even} \\ \Delta t, & n \text{ odd} \end{cases}.$$

Thus, with the aid of the lemma, it follows that for both n odd and even,

$$v_{n,x} = v_{n,x^*} + c.$$

Thus for all $n = 0, 1, 2, 3, \ldots,$

$$F_{n,x^*}^* = F_{n,x} = m \frac{v_{n+1,x^*} + c - v_{n,x^*} - c}{\Delta t} = m \frac{v_{n+1,x^*} - v_{n,x^*}}{\Delta t}. \quad (2.39)$$

Under the assumption that

$$y^* = y - dt_n$$

in which d is a constant, one finds similarly that

$$F_{n,y^*}^* = m \frac{v_{n+1,y^*} - v_{n,y^*}}{\Delta t}, \quad (2.40)$$

and the covariance is established.   □

2.5.   Perihelion Motion

In this section and in the next two sections, we show how to apply conservative methodology to problems in physics. As a first application, let us examine a planar 3-body problem in which the force of interaction is gravitation. In such problems conservation of energy, linear momentum, and angular momentum are basic.

Let $P_i, i = 1, 2, 3$, be three bodies, with respective masses m_i, in motion in the XY plane, in which the force of interaction is gravitation. The force $\vec{F}_{i,j}$ between any two of the bodies will have magnitude $F_{i,j} = G \frac{m_i m_j}{r_{ij}^2}$, in which $G = (6.67)10^{-8}$, and r_{ij} is the distance between the bodies. For initial data, let us choose

$$m_1 = (6.67)^{-1}10^8 \qquad m_2 = (6.67)^{-1}10^6 \qquad m_3 = (6.67)^{-1}10^5$$
$$x_{1,0} = 0.0 \qquad\qquad x_{2,0} = 0.5 \qquad\qquad x_{3,0} = -1.0$$
$$y_{1,0} = 0.0 \qquad\qquad y_{2,0} = 0.0 \qquad\qquad y_{3,0} = 8.0$$
$$v_{1,0,x} = 0.0 \qquad\qquad v_{2,0,x} = 0.0 \qquad\qquad v_{3,0,x} = 0.0$$
$$v_{1,0,y} = 0.0 \qquad\qquad v_{2,0,y} = 1.63 \qquad\qquad v_{3,0,y} = -3.75.$$

The differential equations of motion for this system are

$$m_1\ddot{x}_1 = -\frac{Gm_1m_2}{r_{12}^2}\frac{x_1 - x_2}{r_{12}} - \frac{Gm_1m_3}{r_{13}^2}\frac{x_1 - x_3}{r_{13}}$$

$$m_1\ddot{y}_1 = -\frac{Gm_1m_2}{r_{12}^2}\frac{y_1 - y_2}{r_{12}} - \frac{Gm_1m_3}{r_{13}^2}\frac{y_1 - y_3}{r_{13}}$$

$$m_2\ddot{x}_2 = -\frac{Gm_1m_2}{r_{12}^2}\frac{x_2 - x_1}{r_{12}} - \frac{Gm_2m_3}{r_{23}^2}\frac{x_2 - x_3}{r_{23}}$$

$$m_2\ddot{y}_2 = -\frac{Gm_1m_2}{r_{12}^2}\frac{y_2 - y_1}{r_{12}} - \frac{Gm_2m_3}{r_{23}^2}\frac{y_2 - y_3}{r_{23}}$$

$$m_3\ddot{x}_3 = -\frac{Gm_1m_3}{r_{13}^2}\frac{x_3 - x_1}{r_{13}} - \frac{Gm_2m_3}{r_{23}^2}\frac{x_3 - x_2}{r_{23}}$$

$$m_3\ddot{y}_3 = -\frac{Gm_1m_3}{r_{13}^2}\frac{y_3 - y_1}{r_{13}} - \frac{Gm_2m_3}{r_{23}^2}\frac{y_3 - y_2}{r_{23}},$$

in which

$$r_{ij}^2 = (x_i - x_j)^2 + (y_i - y_j)^2.$$

For $\Delta t = 0.001$, and $i = 1,2,3;\ n = 0,1,2,\ldots$, we approximate the solution of this system with the following form of the recursion formulas, which is most convenient for Newtonian iteration, since the denominators in the iteration formulas are all unity:

$$x_{i,n+1} - x_{i,n} - \frac{1}{2}(\Delta t)(v_{i,n+1,x} + v_{i,n,x}) = 0$$

$$y_{i,n+1} - y_{i,n} - \frac{1}{2}(\Delta t)(v_{i,n+1,y} + v_{i,n,y}) = 0$$

$$v_{i,n+1,x} - v_{i,n,x} - \frac{\Delta t}{m_i}F_{i,n,x} = 0$$

$$v_{i,n+1,y} - v_{i,n,y} - \frac{\Delta t}{m_i}F_{i,n,y} = 0,$$

in which the F's are given by

$$F_{1,n,x} = -\frac{Gm_1m_2[(x_{1,n+1} + x_{1,n}) - (x_{2,n+1} + x_{2,n})]}{r_{12,n}r_{12,n+1}[r_{12,n} + r_{12,n+1}]}$$
$$- \frac{Gm_1m_3[(x_{1,n+1} + x_{1,n}) - (x_{3,n+1} + x_{3,n})]}{r_{13,n}r_{13,n+1}[r_{13,n} + r_{13,n+1}]}$$

$$F_{1,n,y} = -\frac{Gm_1m_2[(y_{1,n+1} + y_{1,n}) - (y_{2,n+1} + y_{2,n})]}{r_{12,n}r_{12,n+1}[r_{12,n} + r_{12,n+1}]}$$
$$- \frac{Gm_1m_3[(y_{1,n+1} + y_{1,n}) - (y_{3,n+1} + y_{3,n})]}{r_{13,n}r_{13,n+1}[r_{13,n} + r_{13,n+1}]}$$

$$F_{2,n,x} = -\frac{Gm_1m_2[(x_{2,n+1} + x_{2,n}) - (x_{1,n+1} + x_{1,n})]}{r_{12,n}r_{12,n+1}[r_{12,n} + r_{12,n+1}]}$$
$$- \frac{Gm_2m_3[(x_{2,n+1} + x_{2,n}) - (x_{3,n+1} + x_{3,n})]}{r_{23,n}r_{23,n+1}[r_{23,n} + r_{23,n+1}]}$$

$$F_{2,n,y} = -\frac{Gm_1m_2[(y_{2,n+1} + y_{2,n}) - (y_{1,n+1} + y_{1,n})]}{r_{12,n}r_{12,n+1}[r_{12,n} + r_{12,n+1}]}$$
$$- \frac{Gm_2m_3[(y_{2,n+1} + y_{2,n}) - (y_{3,n+1} + y_{3,n})]}{r_{23,n}r_{23,n+1}[r_{23,n} + r_{23,n+1}]}$$

$$F_{3,n,x} = -\frac{Gm_1m_3[(x_{3,n+1} + x_{3,n}) - (x_{1,n+1} + x_{1,n})]}{r_{13,n}r_{13,n+1}[r_{13,n} + r_{13,n+1}]}$$
$$- \frac{Gm_2m_3[(x_{3,n+1} + x_{3,n}) - (x_{2,n+1} + x_{2,n})]}{r_{23,n}r_{23,n+1}[r_{23,n} + r_{23,n+1}]}$$

$$F_{3,n,y} = -\frac{Gm_1m_3[(y_{3,n+1} + y_{3,n}) - (y_{1,n+1} + y_{1,n})]}{r_{13,n}r_{13,n+1}[r_{13,n} + r_{13,n+1}]}$$
$$- \frac{Gm_2m_3[(y_{3,n+1} + y_{3,n}) - (y_{2,n+1} + y_{2,n})]}{r_{23,n}r_{23,n+1}[r_{23,n} + r_{23,n+1}]},$$

and

$$r_{ij,m}^2 = (x_{i,m} - x_{j,m})^2 + (y_{i,m} - y_{j,m})^2; \quad m = n, n+1.$$

In the absence of P_3, the motion of P_2 relative to P_1 is the periodic orbit shown in Figure 2.1, for which the period is $\tau = 3.901$. If the major axis of motion is the line of greatest distance between any two points of an orbit, and if the length of the major axis is defined to be $2a$, the major axis of P_2's motion relative to P_1 lies on the X axis and $a = 0.730$.

The initial data for P_3 were chosen so that this body begins its motion relatively far from both P_1 and P_2, arrives in the vicinity of $(-1, 0)$ almost simultaneously with P_2 and then proceeds past $(-1, 0)$ at a relatively

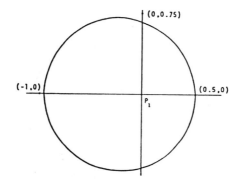

Figure 2.1. A periodic orbit.

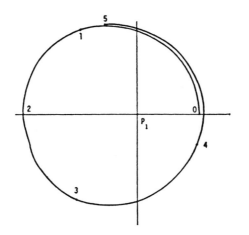

Figure 2.2. Orbit deflection.

high speed, assuring only a short period of strong gravitational attraction. Particles P_2 and P_3 come closest in the third quadrant at t_{2125}, when P_2 is at $(-0.9296, -0.1108)$ and P_3 is at $(-0.9325, -0.1012)$. The effect of the interaction is to deflect P_2 outward, as is seen clearly in Figure 2.2, where the motion of P_2 relative to P_1 has been plotted from t_0 to t_{5000}, with the integer labels $n = 0, 1, 2, 3, 4, 5$, marking the positions t_{1000n}. After having been deflected, P_2 goes into the new orbit about P_1 which is shown in Figure 2.3. The end points of the new major axis are $(0.4943, 0.1664)$ and $(-0.9105, -0.3075)$, so that $a = 0.74135$. The new period is $\tau = 3.9905$.

Now, the *perihelion* point is the position of P_2 which is closest to P_1 during the orbit. Since P_2 has been deflected into a new orbit, its perihelion point has moved. The perihelion motion is measured by the angle

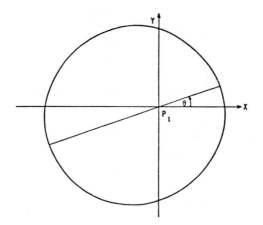

Figure 2.3.   Positive perihelion motion.

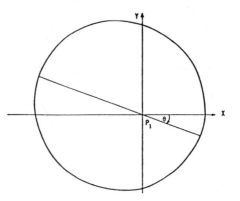

Figure 2.4.   Negative perihelion motion.

of inclination θ of the new major axis with the X axis. and is given by $\tan\theta = 0.34$. The perihelion motion of this example is positive.

If we now change the initial data of P_3 to $x_{3,0} = -0.5$, $y_{3,0} = 8.0$, $v_{3,0,x} = -0.25$, $v_{3,0,y} = -4.00$, then the strongest gravitational effect between P_2 and P_3 occurs in the second quadrant at t_{1966} when P_2 is at $(-0.94582, 0.01950)$ and P_3 is at $(-0.94418, 0.01796)$. P_2 is then perturbed into the new orbit shown in Figure 2.4. The end points of the new major axis are $(0.50724, -0.18349)$ and $(-0.92692, 0.33474)$, so that $a = 0.76246$. The new period is $\tau = 4.162$. The resulting perihelion motion is now negative, since the angle θ of the new major axis with the X axis is given by $\tan\theta = -0.36$.

From the above and similar examples, it follows that the major axis of P_2 is deflected in the same direction as is P_2. In actual planetary motions, for example, in a Sun–Mercury–Venus system, where the mass of the sun is distinctly dominant, it would appear that when Mercury and Venus are relatively close in the first or third quadrants, the perihelion motion of Mercury should be perturbed a small amount in the positive angular direction, while relative closeness in the second or fourth quadrants should result in a small negative angular perturbation. All such possibilities can occur for the motions of Mercury and Venus. Thus, the perihelion motion of Mercury should be a complex, nonlinear, oscillatory motion. This conclusion was verified on the computer with ten full orbits of Mercury.

2.6.   The Fundamental Problem of Electrostatics

The fundamental problem of electrostatics is a conservative problem which is described as follows. Given m electric charges $q_1, q_2, \ldots, q_m$, called the *source charges*, and n electric charges $Q_1, Q_2, \ldots, Q_n$, called the *test charges*, calculate the trajectories of $Q_1, Q_2, \ldots, Q_n$ from given initial data if the positions of the source charges are fixed (Griffiths (1981)). The fundamental problem is a *discrete* problem and has all the inherent difficulties of an n-body problem when $n \geq 3$. The classical way to avoid these difficulties is to consider special classes of problems in which the source charges are distributed continuously, thus allowing the introduction of integrals, fields, Gauss's law, Laplace's equation, and Poisson's equation. In this section we will show how to solve the fundamental problem when m and n are finite. We will use Coulomb's law in the following way. If two particles P_1, P_2 have respective charges e_1, e_2, then a potential ϕ defined by them is taken to be $e_1 e_2 / r_{12}$, in which r_{12} is the distance between them.

For convenience we now let the test charges be $Q_1, Q_2, \ldots, Q_n$ and let the source charges be $q_{n+1}, q_{n+2}, \ldots, q_N$, in which $N = n + m$. Then the motion of the test charges is determined by (2.4), (2.5).

As an example let us consider the following. Let Q_1, Q_2, Q_3 be electrons and let q_1 be a positron which is fixed at the origin $(0, 0, 0)$ of xyz space. The mass of each particle is $(9.1085)10^{-28}$ g. The charge of each of Q_1, Q_2, Q_3 is $-(4.8028)10^{-10}$ esu, while the charge of q_1 is $(4.8028)10^{-10}$ esu. The transformations

$$\boldsymbol{R} = (X, Y, Z) = 10^{12}(x, y, z) = 10^{12}\boldsymbol{r} \tag{2.41}$$

$$T = 10^{22}t \tag{2.42}$$

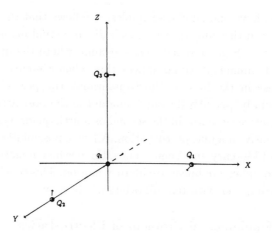

Figure 2.5.   Initial data.

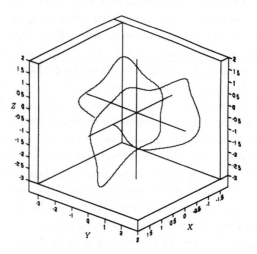

Figure 2.6.   Motion of Q_1.

are introduced for the actual calculations. In the XYZ variables, the
initial positions of Q_1, Q_2, Q_3 are taken to be $(1,0,0), (0,1,0), (0,0,1)$,
respectively. The initial velocities of Q_1, Q_2, Q_3 are taken to be
$(0,1,0), (0,0,1), (1,0,0)$, respectively. These initial data are shown in
Figure 2.5. The initial energy is $-(6.606)10^{-8}$ erg. Finally, let $\Delta T =$
0.00001.

Figures 2.6–2.8 show the complex motions of Q_1, Q_2, Q_3, respectively,
every 5000 time steps over 5000000 time steps. The trajectories are complex

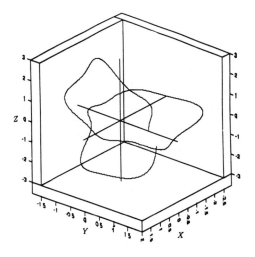

Figure 2.7.   Motion of Q_2.

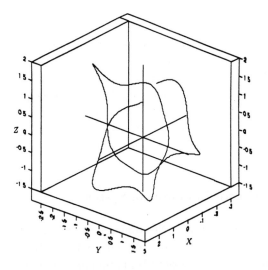

Figure 2.8.   Motion of Q_3.

three dimensional motions, which are not available analytically. In all cases, one finds that $|X_i| < 3, |Y_i| < 3, |Z_i| < 3, i = 1, 2, 3$ and that each X, Y, Z takes on both positive and negative values. The three electrons are usually well separated, as is shown typically in Table 2.1, while the system is held together by the single positron at the origin. The entries in the

Table 2.1. Positions of Q_1, Q_2, Q_3.

T	Q	X	Y	Z
1000000	Q_1	-0.7091	2.4993	0.6711
	Q_2	0.6711	-0.7091	2.4993
	Q_3	2.4993	0.6711	-0.7091
2000000	Q_1	1.2035	-0.9622	-0.2833
	Q_2	-0.2833	1.2035	-0.9622
	Q_3	-0.9622	-0.2833	1.2035
3000000	Q_1	1.6067	1.3750	-0.9647
	Q_2	-0.9647	1.6067	1.3750
	Q_3	1.3750	-0.9647	1.6067
4000000	Q_1	-0.4644	-1.7039	1.4298
	Q_2	1.4297	-0.4644	-1.7039
	Q_3	-1.7039	1.4296	-0.4644
5000000	Q_1	1.5599	-0.2895	0.4383
	Q_2	0.4354	1.5472	-0.2751
	Q_3	-0.2727	0.4340	1.5471

table indicate that, to four decimal places, there may be some symmetry in the three trajectories due to the special initial conditions of the problem. However, this is revealed to be false at the times $T = 4000000, 5000000$. Thereafter, the system becomes physically unstable as Q_2 begins to oscillate around the positron, thus negating the effect of the positron on Q_1 and Q_3. Replacement of the positron by a fixed positive charge three times that of the positron yields a physically stable system with electron trajectories as complex as those shown in Figures 2.6–2.8.

2.7. The Calogero Hamiltonian System

Thus far we have emphasized a Newtonian formulation of the N-body problem. However, a more general formulation would use Hamiltonians. In this section we will discuss a Calogero Hamiltonian system and in the next we will discuss a Toda Hamiltonian system.

A Calogero Hamiltonian system (Calogero (1975), Marsden (1981)) is a system of N particles on a line with Hamiltonian

$$H = \frac{1}{2} \sum_{i=1}^{N} p_i^2 + \sum_{\substack{i \neq j \\ i,j=1}}^{N} \frac{1}{(q_i - q_j)^2}. \tag{2.43}$$

For (2.43),

$$\sum_{i=1}^{N} p_i \tag{2.44}$$

is a system invariant.

We will show how to reformulate particle interactions characterized by (2.43) by means of difference equations so that (2.43) and (2.44) continue to remain system invariants.

For clarity and intuition, let us begin with a two-particle Calogero system whose Hamiltonian is

$$H = \frac{1}{2} \sum_{i=1}^{2} p_i^2 + \frac{2}{(q_1 - q_2)^2}, \quad q_1 \neq q_2. \tag{2.45}$$

Let $\Delta t > 0$, and $t_k = k\Delta t, k = 0, 1, 2, \ldots$. Denote p_1, p_2, q_1, q_2 at time t_k by $p_{1,k}, p_{2,k}, q_{1,k}, q_{2,k}$, respectively. For $i = 1, 2$ and $k = 0, 1, 2, \ldots$, define

$$\frac{p_{i,k+1} + p_{i,k}}{2} = \frac{q_{i,k+1} - q_{i,k}}{\Delta t} \tag{2.46}$$

$$\frac{p_{i,k+1} - p_{i,k}}{\Delta t} = F_{i,k}, \tag{2.47}$$

where

$$F_{1,k} = 2\frac{q_{1,k+1} + q_{1,k} - q_{2,k+1} - q_{2,k}}{(q_{1,k} - q_{2,k})^2 (q_{1,k+1} - q_{2,k+1})^2}, \quad q_{1,k} \neq q_{2,k}, \quad k = 0, 1, 2, \ldots, \tag{2.48}$$

$$F_{2,k} = -F_{1,k}. \tag{2.49}$$

Theorem 2.8.   *Given* $p_{1,0}, p_{2,0}, q_{1,0}, q_{2,0}$, *then* (2.47)–(2.49) *imply the invariance of* $p_{1,k} + p_{2,k}$, *that is,*

$$p_{1,k} + p_{2,k} \equiv p_{1,0} + p_{2,0}, \quad k = 0, 1, 2, \ldots. \tag{2.50}$$

Proof.   From (2.47)–(2.49)

$$p_{1,k+1} = p_{1,k} + (\Delta t)F_{1,k},$$
$$p_{2,k+1} = p_{2,k} - (\Delta t)F_{1,k}.$$

Hence, for $k = 0, 1, 2, \ldots,$

$$p_{1,k+1} + p_{2,k+1} = p_{1,k} + p_{2,k},$$

which implies (2.50). $\qquad\qquad\qquad\qquad\qquad\qquad\qquad\qquad\qquad\square$

Theorem 2.9.   *Difference formulas (2.46)–(2.49) imply the invariance of Hamiltonian (2.45) for given* $p_{1,0}, p_{2,0}, q_{1,0}, q_{2,0}$, *that is, for* $k = 0, 1, 2, \ldots$,

$$\frac{1}{2}\left(p_{1,k}^2 + p_{2,k}^2\right) + \frac{1}{(q_{1,k} - q_{2,k})^2} \equiv \frac{1}{2}\left(p_{1,0}^2 + p_{2,0}^2\right) + \frac{2}{(q_{1,0} - q_{2,0})^2}. \qquad (2.51)$$

Proof.   Let

$$W_n = \sum_{k=0}^{n-1}\left[(q_{1,k+1} - q_{1,k})F_{1,k} + (q_{2,k+1} - q_{2,k})F_{2,k}\right]. \qquad (2.52)$$

Then, from (2.46) and (2.47),

$$
\begin{aligned}
W_n &= \sum_{k=0}^{n-1}\left[(q_{1,k+1} - q_{1,k})\frac{p_{1,k+1} - p_{1,k}}{\Delta t} + (q_{2,k+1} - q_{2,k})\frac{p_{2,k+1} - p_{2,k}}{\Delta t}\right] \\
&= \sum_{k=0}^{n-1}\left[\frac{q_{1,k+1} - q_{1,k}}{\Delta t}(p_{1,k+1} - p_{1,k}) + \frac{q_{2,k+1} - q_{2,k}}{\Delta t}(p_{2,k+1} - p_{2,k})\right] \\
&= \frac{1}{2}\sum_{k=0}^{n-1}\left[(p_{1,k+1}^2 - p_{1,k}^2) + (p_{2,k+1}^2 - p_{2,k}^2)\right]
\end{aligned}
$$

so that

$$W_n = \frac{1}{2}\left(p_{1,n}^2 + p_{2,n}^2\right) - \frac{1}{2}\left(p_{1,0}^2 - p_{2,0}^2\right). \qquad (2.53)$$

However, Eqs. (2.48), (2.49), and (2.52) imply

$$
\begin{aligned}
W_n &= 2\sum_{k=0}^{n-1}\left[(q_{1,k+1} - q_{1,k})\frac{q_{1,k+1} + q_{1,k} - q_{2,k+1} - q_{2,k}}{(q_{1,k} - q_{2,k})^2(q_{1,k+1} - q_{2,k+1})^2}\right. \\
&\qquad\qquad \left. - (q_{2,k+1} - q_{2,k})\frac{q_{1,k+1} + q_{1,k} - q_{2,k+1} - q_{2,k}}{(q_{1,k} - q_{2,k})^2(q_{1,k+1} - q_{2,k+1})^2}\right] \\
&= 2\sum_{k=0}^{n-1}\frac{(q_{1,k+1} - q_{2,k+1})^2 - (q_{1,k} - q_{2,k})^2}{(q_{1,k} - q_{2,k})^2(q_{1,k+1} - q_{2,k+1})^2} \\
&= 2\sum_{k=0}^{n-1}\left[\frac{1}{(q_{1,k} - q_{2,k})^2} - \frac{1}{(q_{1,k+1} - q_{2,k+1})^2}\right],
\end{aligned}
$$

so that

$$W_n = \frac{2}{(q_{1,0} - q_{2,0})^2} - \frac{2}{(q_{1,n} - q_{2,n})^2}. \qquad (2.54)$$

Elimination of W_n between (2.53) and (2.54) then yields (2.51) and the theorem is proved.

The extension to systems of N particles then follows directly from formulation (2.46)–(2.49), but with $i = 1, 2, \ldots, N$.

To implement the formulation practically, consider system (2.46)–(2.49) with the initial data $p_{1,0} = 1, p_{2,0} = -1, q_{1,0} = 1, q_{2,0} = -1$. Then the Newtonian iteration formulas for solving the system at t_{k+1} in terms of data at t_k are

$$q_{1,k+1}^{(n+1)} = q_{1,k} + \Delta t \left[\frac{p_{1,k+1}^{(n)} + p_{1,k}}{2} \right] \tag{2.55}$$

$$q_{2,k+1}^{(n+1)} = q_{2,k} + \Delta t \left[\frac{p_{2,k+1}^{(n)} + p_{2,k}}{2} \right] \tag{2.56}$$

$$p_{1,k+1}^{(n+1)} = p_{1,k} + 2\Delta t \left[\frac{q_{1,k+1}^{(n+1)} + q_{1,k} - q_{2,k+1}^{(n+1)} - q_{2,k}}{(q_{1,k} - q_{2,k})^2 \left(q_{1,k+1}^{(n+1)} - q_{2,k+1}^{(n+1)} \right)^2} \right] \tag{2.57}$$

$$p_{2,k+1}^{(n+1)} = p_{2,k} - 2\Delta t \left[\frac{q_{1,k+1}^{(n+1)} + q_{1,k} - q_{2,k+1}^{(n+1)} - q_{2,k}}{(q_{1,k} - q_{2,k})^2 \left(q_{1,k+1}^{(n+1)} - q_{2,k+1}^{(n+1)} \right)^2} \right]. \tag{2.58}$$

Calculation for 500000 steps with $\Delta t = 0.0001$ yields the typical results shown in Table 2.2 every 50000 time steps. The table shows clearly that both the Hamiltonian and $p_1 + p_2$ are conserved. In addition, it shows an increasingly repulsive effect which the particles exert on each other.   □

Table 2.2. Calogero.

k	H	q_1	q_2	p_1	p_2
1	1.5	1.000000	−1.000000	1.000000	−1.000000
50000	1.5	6.964067	−6.964067	1.220529	−1.220529
100000	1.5	13.076572	−13.076572	1.223551	−1.223551
150000	1.5	19.196231	−19.196231	1.224191	−1.224191
200000	1.5	25.317857	−25.317857	1.224427	−1.224427
250000	1.5	31.440301	−31.440301	1.224539	−1.224539
300000	1.5	37.563163	−37.563163	1.224601	−1.224601
350000	1.5	43.686267	−43.686267	1.224638	−1.224638
400000	1.5	49.809524	−49.809524	1.224663	−1.224663
450000	1.5	55.932884	−55.932884	1.224680	−1.224680
500000	1.5	62.056317	−62.056317	1.224692	−1.224692

2.8. The Toda Hamiltonian System

A Toda Hamiltonian system (Toda (1967)) is a system of N particles on a line with Hamiltonian

$$H = \frac{1}{2}\sum_{i=1}^{N} p_i^2 + \sum_{i=1}^{N-1} \exp(q_i - q_{i+1}).\tag{2.59}$$

We will show how to reformulate particle interactions characterized by (2.59) by means of difference equations so that (2.44) and (2.59) remain system invariants. Formulas (2.46)–(2.49), for $N = 2$, need be modified only slightly, that is, (2.48) needs to be changed, and this is done as follows:

$$F_{1,k} = \begin{cases} -\frac{\exp(q_{1,k+1}-q_{2,k+1})-\exp(q_{1,k}-q_{2,k})}{(q_{1,k+1}-q_{2,k+1})-(q_{1,k}-q_{2,k})}; & (q_{1,k+1} - q_{1,k}) - (q_{2,k+1} - q_{2,k}) \neq 0 \\ -\exp(q_{1,k} - q_{2,k}); & (q_{1,k+1} - q_{1,k}) - (q_{2,k+1} - q_{2,k}) = 0. \end{cases}$$
$$\tag{2.60}$$

Theorem 2.10.   *Given $p_{1,0}, p_{2,0}, q_{1,0}, q_{2,0}$, then Eqs. (2.46), (2.47), (2.49), (2.60) imply*

$$p_{1,k} + p_{2,k} \equiv p_{1,0} + p_{2,0}, \quad k = 0, 1, 2, \ldots.\tag{2.61}$$

Proof.   The proof is essentially identical to that of Theorem 2.8.   □

Theorem 2.11.   *Under the assumptions of Theorem 2.10, it follows for $k = 0, 1, 2, \ldots$, that*

$$\frac{1}{2}\left(p_{1,k}^2 + p_{2,k}^2\right) + \exp(q_{1,k} - q_{2,k}) \equiv \frac{1}{2}\left(p_{1,0}^2 + p_{2,0}^2\right) + \exp(q_{1,0} - q_{2,0}).\tag{2.62}$$

Proof.   Consider first the case $(q_{1,k+1} - q_{1,k}) - (q_{2,k+1} - q_{2,k}) \neq 0$. Recall also Eq. (2.52), that is,

$$W_n = \sum_{k=0}^{n-1}\left[(q_{1,k+1} - q_{1,k})F_{1,k} + (q_{2,k+1} - q_{2,k})F_{2,k}\right].$$

Thus, Eq. (2.53), that is

$$W_n = \frac{1}{2}\left(p_{1,n}^2 + p_{2,n}^2\right) - \frac{1}{2}\left(p_{1,0}^2 - p_{2,0}^2\right)$$

is again valid, using the same argument as used to derive (2.53).

Next, Eqs. (2.46), (2.47), (2.49), (2.52) and (2.60) imply

$$
\begin{aligned}
W_n &= \sum_{k=0}^{n-1} \left\{ (q_{1,k+1} - q_{1,k}) \left[-\frac{\exp(q_{1,k+1} - q_{2,k+1}) - \exp(q_{1,k} - q_{2,k})}{(q_{1,k+1} - q_{2,k+1}) - (q_{1,k} - q_{2,k})} \right] \right. \\
&\left. \quad + (q_{2,k+1} - q_{2,k}) \left[\frac{\exp(q_{1,k+1} - q_{2,k+1}) - \exp(q_{1,k} - q_{2,k})}{(q_{1,k+1} - q_{2,k+1}) - (q_{1,k} - q_{2,k})} \right] \right\} \\
&= \sum_{k=0}^{n-1} \{ -[\exp(q_{1,k+1} - q_{2,k+1}) - \exp(q_{1,k} - q_{2,k})] \},
\end{aligned}
$$

so that

$$
W_n = \exp(q_{1,0} - q_{2,0}) - \exp(q_{1,n} - q_{2,n}). \tag{2.63}
$$

Finally, elimination of W_n between Eqs. (2.53) and (2.63) yields (2.62).

In the second case, when $(q_{1,k+1} - q_{1,k}) - (q_{2,k+1} - q_{2,k}) = 0$, the corresponding summation term in (2.52) becomes simply

$$
[(q_{1,k+1} - q_{1,k}) - (q_{2,k+1} - q_{2,k})][-\exp(q_{1,k} - q_{2,k})]
$$

which is zero, and the theorem continues to be valid. $\qquad \square$

Practical implementation uses formulas entirely analogous to (2.55)–(2.58), but which incorporate (2.60) for the Toda lattice. Calculation for 240000 steps with $\Delta t = 0.000001$ with initial data $q_{1,0} = 1$, $q_{2,0} = -1$, $p_{1,0} = 10$, $p_{2,0} = -10$ yields the results in Table 2.3. The second part of (2.60) is essential numerically at the turning point, which occurs between

Table 2.3.   Toda.

k	H	q_1	q_2	p_1	p_2
1	107.39	1.000000	-1.000000	10.000000	-10.000000
20000	107.39	1.198295	-1.198295	9.818524	-9.818524
40000	107.39	1.392152	-1.392152	9.549895	-9.549895
60000	107.39	1.579463	-1.579463	9.156625	-9.156625
80000	107.39	1.757265	-1.757265	8.590050	-8.590050
100000	107.39	1.921525	-1.921525	7.792399	-7.792399
120000	107.39	2.067027	-2.067027	6.7051120	-6.7051120
140000	107.39	2.187513	-2.187513	5.286525	-5.286525
160000	107.39	2.276273	-2.276273	3.537755	-3.537755
180000	107.39	2.327260	-2.327260	1.526700	-1.526700
200000	107.39	2.336502	-2.336502	-0.609200	0.609200
220000	107.39	2.303235	-2.303235	-2.694399	2.694399
240000	107.39	2.230142	-2.230142	-4.569209	4.569209

$k = 180000$ and $k = 200000$. The table indicates clearly the invariance of both H and $p_1 + p_2$.

2.9.   Remarks

In applying Newton's iteration formulas to the resulting algebraic or transcendental system of the method of Section 2.2, it is convenient to know how many solutions the system has. We now give an example to show that the solution need not be unique, and indeed has two solutions. Each problem one considers will require a related analysis.

Consider the initial value problem

$$\ddot{x} = x^2, \quad x(0) = 1, \quad \dot{x} = 1. \tag{2.64}$$

Choosing $\phi(x) = -\frac{1}{3}x^3$, the system to be solved is

$$x_{k+1} = x_k + \frac{1}{2}(\Delta t)(v_{k+1} + v_k) \tag{2.65}$$

$$v_{k+1} = v_k + \frac{1}{3}(\Delta t)\left(x_{k+1}^2 + x_{k+1}x_k + x_k^2\right). \tag{2.66}$$

Substitution of (2.66) into (2.65) yields

$$x_{k+1}^2 + \left(1 - \frac{6}{(\Delta t)^2}\right)x_{k+1} + \left(1 + \frac{6}{(\Delta t)^2} + \frac{6}{(\Delta t)}\right) = 0. \tag{2.67}$$

Since the initial conditions are given in (2.38), it follows from (2.67) that

$$x_1^2 + \left(1 - \frac{6}{(\Delta t)^2}\right)x_1 + \left(1 + \frac{6}{(\Delta t)^2} + \frac{6}{(\Delta t)}\right) = 0. \tag{2.68}$$

However, examination of the discriminant of (2.68) reveals that for $\Delta t < 0.79490525$, Eq. (2.68) has two real roots. Indeed, one must choose the negative sign in the quadratic formula to get the correct root. For $\Delta t = 0.01$, the correct physical approximation is $x_1 = 1.01005$, while the incorrect solution is $x_1 = 59998$.

Note also that if in (2.5) one would find that any $r_{lm,n+1} = r_{lm,n}$, then $\phi(r_{lm,n+1}) = \phi(r_{lm,n})$. In this case, the corresponding term in (2.6) need only be replaced by

$$-(r_{lm,n+1} - r_{lm,n}) \left. \frac{\partial \phi}{\partial r} \right|_{r=r_{lm,n}}$$

and the theorem will continue to be valid.

Note finally that conservative methodology will be applied again in Chapter 8.

Chapter 3

N-Body Problems with $100 < N \leq 10000$

3.1. Classical Molecular Forces

In this section we will study classical molecular models. We begin then with a review of fundamental concepts and formulas.

From a classical molecular point of view (Feynman, Leighton and Sands (1963)), close molecules behave in the following qualitative fashion:

(a) when pulled apart, they attract;
(b) when pushed together, they repel; and
(c) repulsion is of a greater order of magnitude than is attraction.

The most important exception to this general behavior is the basic fluid of living matter, i.e., liquid water, the primary reason being that close liquid water molecules exhibit hydrogen bonding. However, the number of such bonds, that is, the average cluster size, decreases with pressure and temperature, and there does not seem to be evidence for or against such bonding during turbulent flow, which flow will be emphasized later.

There are a variety of classical molecular potentials for the interactions of molecules and from these classical molecular force formulas can be derived (Hirschfelder, Curtiss and Bird (1967)). The potential which has received the most attention is the Lennard–Jones potential, that is,

$$\phi(r_{ij}) = 4\epsilon \left[\frac{\sigma^{12}}{r_{ij}^{12}} - \frac{\sigma^6}{r_{ij}^6} \right] \text{erg}, \tag{3.1}$$

in which r_{ij} is measured in angstroms. In (3.1), the term with exponents 12 is the repulsion term and the term with exponents 6 is the attraction term. As r_{ij} goes to zero, the resulting motion is volatile.

As an example, an approximate Lennard–Jones potential for water vapor molecules is

$$\phi(r_{ij}) = (1.9646)10^{-13}\left[\frac{2.725^{12}}{r_{ij}^{12}} - \frac{2.725^{6}}{r_{ij}^{6}}\right] \text{ erg } \quad \left(\frac{\text{grcm}^2}{\text{sec}^2}\right). \qquad (3.2)$$

The force $\vec{F}_{ij}$ exerted on P_i by P_j is then

$$\vec{F}_{ij} = (1.9646)10^{-5}\left[\frac{12(2.725^{12})}{r_{ij}^{13}} - \frac{6(2.725^{6})}{r_{ij}^{7}}\right]\frac{\vec{r}_{ji}}{r_{ij}} \text{ dynes } \quad \left(\frac{\text{grcm}}{\text{sec}^2}\right).$$

$$(3.3)$$

Note that $F_{ij} = \|\vec{F}_{ij}\| = 0$ implies that $r_{ij} = 3.06\,\text{Å}$, which is the equilibrium distance, with repulsion prevailing if $r_{ij} < 3.06$ and attraction prevailing if $r_{ij} > 3.06$. Note also that in a regular triangle with edge $r_{ij} = 3.06\,\text{Å}$, the altitude is $2.65\,\text{Å}$.

3.2.   Equations of Motion for Water Vapor Molecules

In molecular mechanics we simulate the motion of a system of molecules using classical molecular potentials and Newtonian mechanics. Thus the motion for a single water vapor molecule P_i acted on by a single water vapor molecule $P_j, i \neq j$, is then

$$m_i\vec{a}_i = (1.9646)10^{-5}\left[\frac{12(2.725^{12})}{r_{ij}^{13}} - \frac{6(2.725^{6})}{r_{ij}^{7}}\right]\frac{\vec{r}_{ji}}{r_{ij}}. \qquad (3.4)$$

Since the mass of a water molecule is $(30.103)10^{-24}$ gr, Eq. (3.4) is equivalent to

$$\vec{a}_i = (160.33)10^{19}\left[\frac{818.90}{r_{ij}^{13}} - \frac{1}{r_{ij}^{7}}\right]\frac{\vec{r}_{ji}}{r_{ij}} \quad \left(\frac{\text{cm}}{\text{sec}^2}\right). \qquad (3.5)$$

Recasting the latter equation in $\text{Å}/(\text{ps}^2)$ yields

$$\vec{a}_i = (160330)\left[\frac{818.90}{r_{ij}^{13}} - \frac{1}{r_{ij}^{7}}\right]\frac{\vec{r}_{ji}}{r_{ij}} \quad \left(\frac{\text{Å}}{\text{ps}^2}\right). \qquad (3.6)$$

On the molecular level, however, the effective force on P_i is local, that is it is determined only by close molecules. But, because of our interest in turbulence, in which the forces of repulsion are so great that attraction is of lesser importance, we will take *local* to mean the solution of the equation

$$\frac{dF_{ij}}{dr_{ij}} = 0. \tag{3.7}$$

The solution to this equation yields $r_{ij} = 3.39\,\text{Å}$. Thus, for $r_{ij} \geq D = 3.39\,\text{Å}$, we choose $\vec{F}_{ij} = \vec{0}$. Note that it is more common to choose $D = 3\sigma$, 4σ, or 5σ.

From (3.6), then, the dynamical equation for water vapor molecule P_i will be

$$\frac{d^2\vec{r}_i}{dt^2} = (160330) \sum_{\substack{j \\ j \neq 1}} \left[\frac{818.90}{r_{ij}^{13}} - \frac{1}{r_{ij}^{7}} \right] \frac{\vec{r}_{ji}}{r_{ij}}; \quad r_{ij} < D. \tag{3.8}$$

The equations of motion for a system of water vapor molecules are then

$$\frac{d^2\vec{r}_i}{dt^2} = (160330) \sum_{\substack{j \\ j \neq 1}} \left[\frac{818.90}{r_{ij}^{13}} - \frac{1}{r_{ij}^{7}} \right] \frac{\vec{r}_{ji}}{r_{ij}}; \quad i = 1, 2, 3, \ldots, N; \quad r_{ij} < D.$$

$$\tag{3.9}$$

Note that on the molecular level gravity can be neglected since $980\,\text{cm/sec}^2 = (980)10^{-16}\,\text{Å/ps}^2$. Note also that in simulating the motion of a system of N molecules, nonlinear equations (3.9) forewarn us that we will have to solve *large* systems of nonlinear, second order, ordinary differential equations (2N in two dimensions and 3N in three dimensions), and that these will have to be solved numerically. The method to be used is based on the leap frog formulas (see Appendix III).

3.3. A Cavity Problem

Let us first construct a regular triangular grid of edge length $3.06\,\text{Å}$, and with altitude $2.65\,\text{Å}$. We determine 4235 grid points in the XY plane as follows:

$$x(1) = -91.8, \quad y(1) = 0,$$
$$x(i) = x(i-1) + 3.06, \quad y(i) = 0, \qquad\qquad i = 2, 61$$
$$x(62) = -90.27, \quad y(62) = 2.65$$
$$x(i) = x(i-1) + 3.06, \quad y(i) = 2.65, \qquad\qquad i = 62, 121$$
$$x(i) = x(i-121), \quad y(i) = y(i-121) + 5.30, \quad i = 122, 4235$$

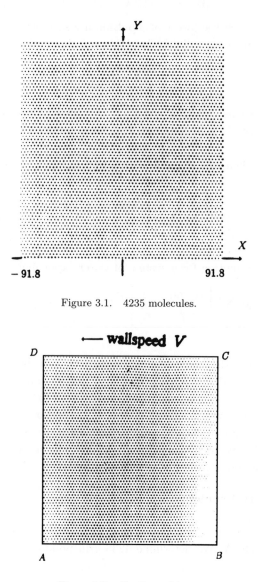

Figure 3.1.   4235 molecules.

Figure 3.2.   Cavity problem.

At each point $(x(i), y(i))$ we set a water vapor molecule $P_i, i = 1, 4235$. This array is shown in Figure 3.1.

   To complete the initial data, note that in two dimensions the rms velocity v for a water vapor molecule at $150°$C is $v = 6.23$ Å/ps. Each molecule is now assigned this speed in either the X or Y direction, determined at random, with its ($\pm$) sign also determined at random.

Now, in two dimensions consider the square of side 183.6 Å, shown in Figure 3.2. The interior is called the cavity or the basin of the square. This basin encloses the molecular fluid shown in Figure 3.1. The four sides are called the walls. The top wall alone, CD, is allowed to move. It moves in the X direction at a constant speed V, called the wallspeed. Also it is allowed an extended length so that the molecular fluid is always completely enclosed by four walls. Then the cavity problem is to describe the gross motion of the fluid for various choices of V, which will be given in Å/ps.

For laboratory studies of the cavity problem see the treadmill apparatus of Freitas, Street, Findikakis and Koseff (1985).

3.4. Computational Considerations

In all of our examples, the following computational considerations are implemented. For time step Δt (ps), and $t_k = k\Delta t, k = 0, 1, 2, \ldots$, two problems must be considered relative to the computations. The first problem is to prescribe a protocol when, computationally, a molecule has crossed a wall into the exterior of the cavity. For each of the lower three walls, we will proceed as follows (no slip condition). The position will be reflected back symmetrically, relative to the wall, into the interior of the basin, the velocity component tangent to the wall will be set to zero and the velocity component perpendicular to the wall will be multiplied by -1. If the molecule has crossed the moving wall, then its position will be reflected back symmetrically, its Y component of velocity will be multiplied by -1, and its X component of velocity will be increased by the wallspeed V.

The second problem derives from the fact that an instantaneous velocity field for molecular motion is Brownian. In order to better interpret gross fluid motion, we will introduce average velocities as follows. For J a positive integer, let particle P_i be at $(x(i, k), y(i, k))$ at t_k and at $(x(i, k - J), y(i, k - J))$ at t_{k-J}. Then the average velocity $\vec{v}_{i,k,J}$ of P_i at t_k is defined by

$$\vec{v}_{i,k,J} = \left(\frac{x(i, k) - x(i, k - J)}{J\Delta t}, \frac{y(i, k) - y(i, k - J)}{J\Delta t} \right). \tag{3.10}$$

In the examples to be described, we will discuss results for various values of J.

Observe also that the connection between the wall speed V and the Reynolds number Re is given (Pan and Acrivos (1967)) by

$$Re = |V|B/\nu, \tag{3.11}$$

in which B is the span, that is, the length CD in Figure 3.2, and ν is the average kinematic viscosity of water.

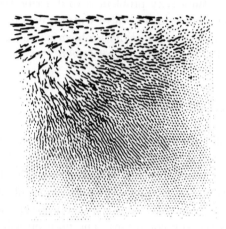

Figure 3.3.   Wallspeed $= -24, J = 20000, t = 4.0$.

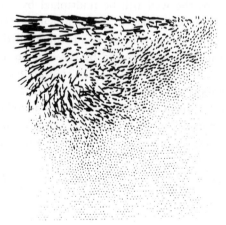

Figure 3.4.   Wallspeed $= -24, J = 20000, t = 8.0$.

3.5. Primary Vortex Generation

Consider first the parameter choices $V = -24\,\text{Å/ps}, J = 20000, \Delta t = 0.0001\,\text{ps}$. Figures 3.3–3.6 show the development of a primary vortex at the respective times $t = 4, 8, 14, 20$. An area of compression is evident in

Figure 3.5.   Wallspeed $= -24, J = 20000, t = 14.0$.

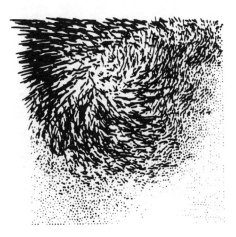

Figure 3.6.   Wallspeed $= -24, J = 20000, t = 20.0$.

Figure 3.3. Other values of J which were studied were 16000, 12000, 9000, and 6000, each of which yielded results similar to those of Figures 3.3–3.6. Note that these results agree qualitatively with experimental results for cavity flow in the large (Freitas *et al.* (1985)).

Results similar to those in Figures 3.3–3.6, but with larger primary vortices, were obtained with $V = -40, -130$, and -600.

3.6.   Turbulent Flow Generation

Turbulence is the most common yet least understood type of fluid flow. Turbulent flows have two well defined characteristics: (1) Many small vortices appear and disappear quickly (Kolmogorov (1964)), and (2) A strong current develops across the usual primary direction (Schlichting (1960)). Though mathematical fluid dynamicists are aware that the Navier–Stokes equations are not the equations of turbulent flow, engineers continue to generate "turbulent" flows using the Navier–Stokes equations with high Reynolds number (Ladyzhenskaya (1969)). It should also be observed that the methods and formulas of statistical mechanics are not applicable to turbulent flow (Bachelor (1959), Bernard (1998)).

The discussion in Section 3.5 for primary vortex generation now leads to the following approach to generating turbulent flows. For a sufficiently large magnitude of the wallspeed V, let us show that turbulence results

Figure 3.7.   Wallspeed $= -3000, J = 80000, t = 0.8$.

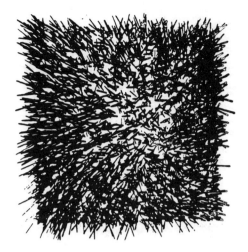

Figure 3.8.   Wallspeed $= -3000, J = 60000, t = 0.8$.

Figure 3.9.   Wallspeed $= -3000, J = 40000, t = 0.8$.

when, for given Δt, a stable calculation results but no J exists which yields a primary vortex.

Let us then set $V = -3000, \Delta t = 2.5(10)^{-6}$. The motion was simulated to $t = 0.8$. Typical results are shown in Figures 3.7–3.10 for

Figure 3.10.   Wallspeed $= -3000, J = 20000, t = 0.8$.

Figure 3.11.   Blowup of a right section of Figure 3.8.

$J = 80000, 60000, 40000, 20000$. None of these figures shows a primary vortex because a strong current has developed across the primary vortex direction. Figure 3.11 shows an enlargement of the section of Figure 3.8 in the range $45 \leq x \leq 91.8, 10 \leq y \leq 100$, and reveals this crosscurrent clearly.

To support the contention that Figures 3.7–3.10 represent fully turbulent flow, we now define the concept of a small vortex. For $3 \leq M \leq 6$,

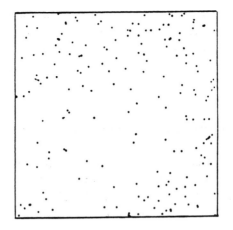

Figure 3.12. 185 small vortices at $t = 0.8$.

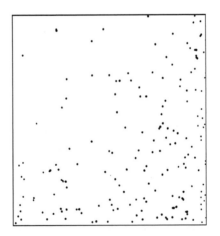

Figure 3.13. 182 small vortices at $t = 0.6$.

we define a *small vortex* as a flow in which M molecules nearest to an $(M + 1)$st molecule rotate either clockwise or counterclockwise about the $(M + 1)$st molecule and, in addition, the $(M + 1)$st molecule lies *interior* to a simple polygon determined by the given M molecules. With this definition, Figure 3.12 shows that the flow in Figure 3.8 has 185 small vortices at $t = 0.8$, while Figure 3.13 shows that only 0.2 picoseconds before, that is at $t = 0.6$, the resulting flow had 182 small vortices which are completely different than those in Figure 3.8.

Primary vortex generation and turbulence have also been studied for the three dimensional cavity problem. New forces not found in 2D arise in a natural way. However, we are limiting our presentations only to personal computer usage, and these three dimensional studies require the use of a supercomputer.

3.7. Primary Vortices and Turbulence for Air

Results entirely analogous to those for water vapor were obtained for the cavity problem for air. But in this case a new problem had to be resolved first. It is rather interesting that even though air is heterogeneous and consists of a variety of atoms and molecules, experimental Lennard–Jones potentials are readily available only for homogeneous air (Hirschfelder, Curtiss and Bird (1967)).

One such potential is

$$\phi(r_{ij}) = (5.36)10^{-14} \left[\frac{3.617^{12}}{r_{ij}^{12}} - \frac{3.617^6}{r_{ij}^6} \right] \text{erg}. \tag{3.12}$$

Before one can proceed to dynamical considerations, it is necessary to characterize carefully the hypothetical air molecule to be used. We assume that the air to be used is *nondilute* and *dry*. Dry air consists primarily of 78% N_2, 21% O_2, and 1% Ar, whose respective masses are

$$m(N_2) = 28(1.660)10^{-24} \text{ gr}$$
$$m(O_2) = 32(1.660)10^{-24} \text{ gr}$$
$$m(Ar) = 40(1.660)10^{-24} \text{ gr}.$$

We now characterize an "air" molecule A as consisting of proportionate amounts of N_2, O_2, and Ar and having mass

$$m(A) = [0.78(28) + 0.21(32) + 0.01(40)](1.660)10^{-24}$$
$$= (4.807)10^{-23} \text{ gr}. \tag{3.13}$$

With this hypothetical air molecule, the computations proceeded as for water vapor.

For the parameter choice $V = -40$, the resulting primary vortex at $t = 10.2$ is slightly larger than the corresponding one obtained for water vapor with $V = -40$ at $t = 10.2$. For the parameter choice $V = -3000$, the turbulent flow is entirely analogous to that for water vapor. The average speed of the air molecules is $362\,\text{Å/ps}$ at $t = 0.6$ while that for water at the same time was $353\,\text{Å/ps}$.

3.8.   Simulation of Cracks and Fractures in a Sheet of Ice

Again we utilize (3.9) and for convenience recall it as

$$\frac{d^2\vec{r}_i}{dt^2} = (160330) \sum_{\substack{j \\ j\neq 1}} \left[\frac{818.90}{r_{ij}^{13}} - \frac{1}{r_{ij}^7} \right] \frac{\vec{r}_{ji}}{r_{ij}}; \quad i = 1, 2, 3, \ldots, N. \qquad (3.14)$$

However we will want to apply it this time to a solid, specifically, to ice, and this will be implemented as follows.

To simulate a rectangular plate of ice molecules, let the points P_i with respective coordinates $(x_i, y_i), i = 1, 2, \ldots, 2713$ be defined by

$$
\begin{aligned}
x(1) &= -59.67, & y(1) &= 0 \\
x(41) &= -61.2, & y(41) &= 2.65 \\
x(i+1) &= x(i) + 3.06, & y(i+1) &= y(1), & i &= 1, 2, \ldots, 39 \\
x(i+1) &= x(i) + 3.06, & y(i+1) &= y(41), & i &= 41, 42, \ldots, 80 \\
x(i) &= x(i - 81), & y(i) &= y(i - 81) + 2(2.65), & i &= 82, 83, \ldots, 2713.
\end{aligned}
$$

At each point (x_i, y_i) set a molecule P_i. The molecular configuration is shown in Figure 3.21. To assure the solidity of the configuration, all initial velocities are set to zero. The neighbors of any P_i are taken to be those molecules which are initially $3.06\,\text{Å}$ from P_i. Moreover, since we will be dealing with a solid, the neighbors of each P_i are defined to be the neighbors of P_i for all time and the force on each P_i is the sum of the forces exerted on P_i by its neighbors.

To facilitate the computations, we now make the transformation $T = 100\,t$, so that (3.14) simplifies to

$$\frac{d^2\vec{r}_i}{dT^2} = (16.0330) \sum_{\substack{j \\ j\neq 1}} \left[\frac{818.90}{r_{ij}^{13}} - \frac{1}{r_{ij}^7} \right] \frac{\vec{r}_{ji}}{r_{ij}}; \quad i = 1, 2, 3, \ldots, 2713. \qquad (3.15)$$

From (3.3) it follows that the elastic limit, found by differentiating F_{ij} and setting the result equal to zero is $r_{ij} = 3.39\,\text{Å}$.

In the examples which follow, system (3.15) is solved by the leap frog formulas with $\Delta T = 0.0001$ and $T_k = k\Delta T, k = 0, 1, 2, \ldots$.

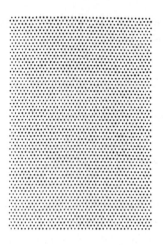

Figure 3.14.   The initial configuration.

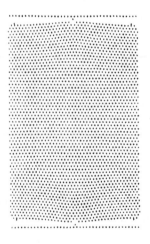

Figure 3.15.   Impulsive reaction.

For our first example, the ice sheet in Figure 3.14 is stressed as follows. At each time step the very top row is moved upward dÅ and the very bottom row is moved downward dÅ. The remaining molecules must respond by (3.15) to this imposed stress.

For $d = 0.000001$, Figure 3.15 shows that the force is impulsive and simply tears off the topmost and bottommost sections of the sheet. Reducing d to 0.0000004 yields the results shown in Figures 3.16–3.18 at the respective times $T(8000000), T(15000000), T(30000000)$. The force field in Figure 3.19 is that for Figure 3.16 and reveals that the force is transmitted internally

Figure 3.16. $T = T(8000000)$.

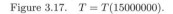

Figure 3.17. $T = T(15000000)$.

in waves. Figure 3.20 shows the force field for Figure 3.17 and indicates the directions of the cracking, which culminate in the fracture shown in Figure 3.18. The present results demonstrate a strong cohesiveness of the sheet, which will be verified by the next example.

Consider now placing a slot in Figure 3.14 by removing the 15 molecules $P(1070 + 41i), i = 0, 14$. The resulting sheet is shown in Figure 3.21. We expect now that the sheet has been weakened. Repeating the stress induced by $d = 0.000001$ no longer results in the impulsive motion shown in Figure 3.15. The sheets reaction at times $T(3000000), T(6000000), T(9000000), T(13000000), T(17000000)$ are

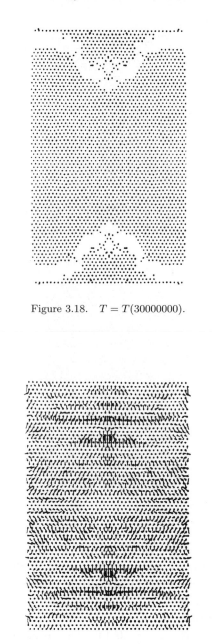

Figure 3.18. $T = T(30000000)$.

Figure 3.19.   Force field of Figure 3.16.

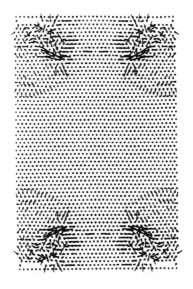

Figure 3.20.   Force field of Figure 3.17.

Figure 3.21.   A 60° slot.

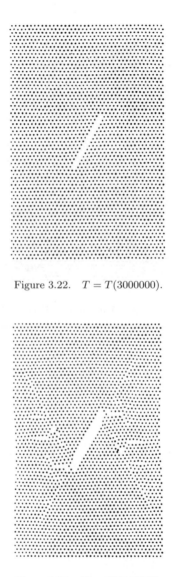

Figure 3.22. $T = T(3000000)$.

Figure 3.23. $T = T(6000000)$.

shown in Figures 3.22–3.26, respectively. The force fields for Figures 3.22 and 3.23 are shown in Figures 3.27 and 3.28, which indicate how the crack patterns are developing and where the final fracture will result. Figure 3.29 further delineates aspects of Figure 3.28 by showing slash lines between those molecules of the original configuration (Figure 3.21) whose separation

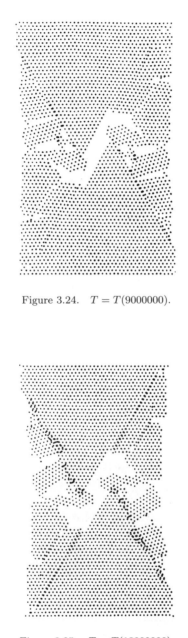

Figure 3.24. $T = T(9000000)$.

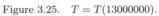

Figure 3.25. $T = T(13000000)$.

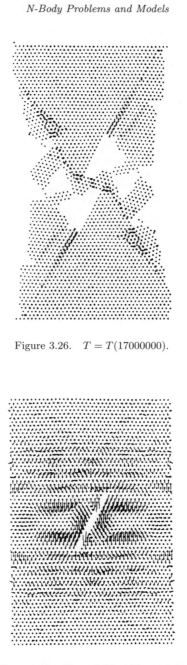

Figure 3.26. $T = T(17000000)$.

Figure 3.27.   Force field of Figure 3.22.

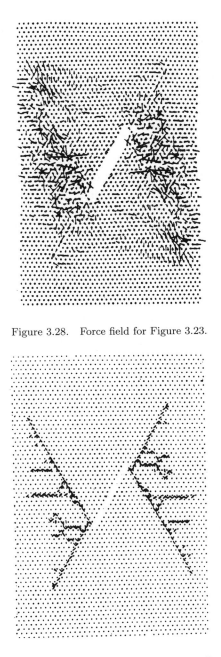

Figure 3.28.   Force field for Figure 3.23.

Figure 3.29.   Molecular separations and cracks in Figure 3.28.

distances have exceeded the elastic limit. In this figure, slash lines which are not connected indicate a weakening of the cohesiveness of the sheet. The connected slash lines represent cracks and these have developed from previously unconnected, but close, slash lines.

3.9.   Shocks in a Nanotube

It is known that on the nano scale, the physics one encounters may be different than that in the large. We explore the general character of such possible differences in this section by simulating conservative and nonconservative shock generation in a nanotube. The gas we will study is air at $20°$C. The mass and potential are those in Section 3.7. Thus,

$$\phi(r_{ij}) = 4\epsilon \left[\frac{\sigma^{12}}{r_{ij}^{12}} - \frac{\sigma^6}{r_{ij}^6} \right] \text{erg}, \tag{3.16}$$

in which $4\epsilon = (5.36)10^{-14}$ and $\sigma = 3.617\,\text{Å}$. From (3.16) the force $\vec{F}_{ij}$ on molecule P_i due to molecule P_j which is r_{ij} Å from P_i is given by

$$\vec{F}_{ij} = (5.36)10^{-6} \left[\frac{12(3.617)^{12}}{r_{ij}^{13}} - \frac{6(3.617)^6}{r_{ij}^7} \right] \frac{\vec{r}_{ij}}{r_{ij}} \text{dynes}. \tag{3.17}$$

The magnitude F_{ij} of $\vec{F}_{ij}$ is

$$F_{ij} = (5.36)10^{-6} \left[\frac{12(3.617)^{12}}{r_{ij}^{13}} - \frac{6(3.617)^6}{r_{ij}^7} \right].$$

Note that $F_{ij} = 0$ implies $r_{ij} = 4.06\,\text{Å}$, which is the equilibrium distance. The effective force on P_i is local so that only those molecules within a distance D, determined by $D = 2.5\sigma = 9\,\text{Å}$, will be considered. Thus, for $r_{ij} \geq D = 9\,\text{Å}$, we choose $\vec{F}_{ij} = 0$. Recall also that

$$m(\text{A}) = [0.78(28) + 0.21(32) + 0.01(40)](1.660)10^{-24} = (4.807)10^{-23} \text{gr}. \tag{3.18}$$

From (3.17) and (3.18), it follows readily that the acceleration, in Å/ps^2, of an air molecule P_i due to interaction with an air molecule P_j satisfies

the equation

$$\frac{d^2\vec{r}_i}{dt^2} = (149795) \left[\frac{4478}{r_{ij}^{13}} - \frac{1}{r_{ij}^7} \right] \frac{\vec{r}_{ji}}{r_{ij}} \text{ Å/ps}^2; \quad r_{ij} < D. \tag{3.19}$$

Observe also that in 2D and at 20°C, the rms speed of an air molecule is 4.103 Å/ps.

Consider now a rectangular nanotube ABCD which is 81 Å by 808 Å, as shown in Figure 3.30 and in which the left wall AD also represents the head of a piston. In the tube we generate a regular triangular grid of 4788 points with fundamental edge length 4.06 Å by the recursion formulas

$$x(1) = 0.0, \quad y(1) = 0.0$$
$$x(i) = x(i-1) + 4.06, \quad y(i) = 0.0, \qquad i = 2, 200$$
$$x(201) = 2.03, \quad y(201) = 3.516$$
$$x(i) = x(i-1) + 4.06, \quad y(i) = 3.516, \qquad i = 202, 399$$
$$x(i) = x(i-399), \quad y(i) = y(i-399) + 7.032, \quad i = 400, 4788.$$

At each point (x_i, y_i) we place an air molecule $P_i, i = 1, 4788$. Each air molecule is assigned a speed of 4.103 Å/ps in the $\pm x$ or $\pm y$ direction, the direction and the sign being determined at random. Then, using the leap frog formulas, the motion of the entire system is simulated for 720000 time steps with $\Delta t = (4)10^{-5}$ with no wall damping and with symmetric reflection from the walls. The resulting configuration is shown in Figure 3.31 and it is used as the initial configuration for all the examples which follow.

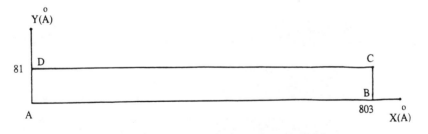

Figure 3.30. A micro tube.

Figure 3.31. Molecular configuration after 720000 steps.

In each of the examples to be discussed, the leap frog formulas will be applied with $\Delta t = 10^{-5}$ ps. Also note that nonconservation will be modelled by the damping of a molecule's velocity if it collides with a wall or the piston head.

Example 1.   Consider first the movement of the piston head into the tube. This is accomplished by increasing the x coordinate of the piston head 0.0004 Å per time step. Thus the piston is moving into the tube at 40 Å/ps. If and when a molecule collides with a wall or with the piston head, it is reflected back into the tube symmetrically with no damping of the velocity. The resulting shock development is shown in Figure 3.32(a)–(f)

Figure 3.32. $V = 40$, no damping.

at the respective times $t = 0.5, 1.0, 1.5, 2.0, 2.5, 3.0, 3.5, 4.0$. The motion is conservative and entirely similar to what one expects in the large.

Example 2. All the considerations of Example 1 are repeated with one exception. When reflecting the molecular path from a wall, the velocity component parallel to the wall is set to zero, while reflecting the molecular path from the piston head, the velocity component parallel to the head is set to zero (*noslip* condition). The results are shown in Figure 3.33(a)–(f) at the respective times $t = 0.5, 1.0, 1.5, 2.0, 2.5, 3.0, 3.5, 4.0$. The results are

Figure 3.33. $V = 40$, parallel damping.

entirely similar to those in Figure 3.32, with the shock fronts moving at
approximately the same speeds. Typical differences can be seen by comparing the upper left corners of Figures 3.32(c) and 3.33(c), and by comparing
the shock fronts in Figures 3.32(d) and 3.33(d).

Example 3.   All the considerations of Example 2 are repeated with
one exception. In addition, the reflected velocity component perpendicular to a wall or to the piston head is now decreased by the factor
0.6. The results are shown in Figure 3.34(a)–(f) at the respective times
$t = 0.5, 1.0, 1.5, 2.0, 2.5, 3.0$. The results now differ in two important ways
from the results in Figures 3.32 and 3.33. First, the shock front has increased
its speed. Second, an area of rarefaction has developed behind the shock
front.

Example 4.   All considerations are the same as in Example 1 with
one exception. The 0.0004 is replaced by 0.0006. Thus the piston moves
into the tube at $60 \, \text{Å/ps}$. The resulting shock development is shown in

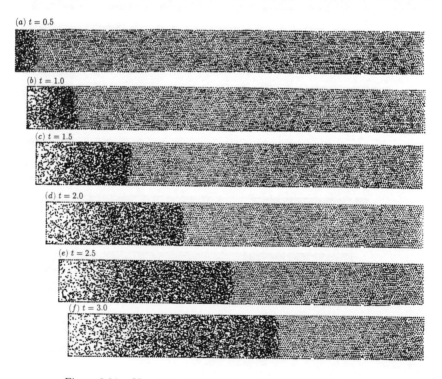

Figure 3.34. $V = 40$, parallel and perpendicular (0.6) damping.

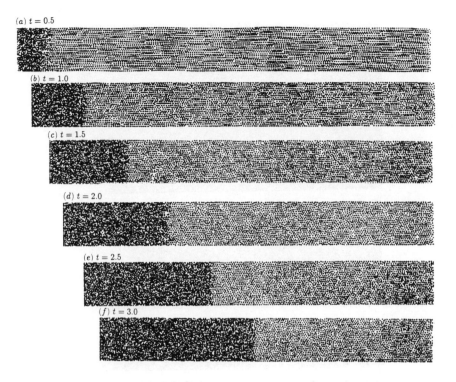

(a) $t = 0.5$

(b) $t = 1.0$

(c) $t = 1.5$

(d) $t = 2.0$

(e) $t = 2.5$

(f) $t = 3.0$

Figure 3.35. $V = 60$, no damping.

Figure 3.35(a)–(f) at the respective times $t = 0.5, 1.0, 1.5, 2.0, 2.5, 3.0$. The motion is entirely similar to what one expects in the large, with speed of the shock front greater by about 1.5 that of Example 1. It is important to note that there is no velocity damping in the simulation.

Example 5. Example 4 is repeated but with two fundamental changes. This time each molecule has its initial x component of velocity decreased by -60 Å/ps and the piston head is kept fixed. The resulting shock motion relative to the piston head shown in Figure 3.36 is entirely similar to, but not exactly the same, as that shown in Figure 3.35. Typical small differences can be seen by comparing the top left particles at $t = 0.5$ and the shock fronts at $t = 3.0$.

Example 6. Example 5 is repeated but with one change. When a molecule is reflected from a wall, its velocity component parallel to the wall or the piston head is set to zero. The resulting shock development is shown in Figure 3.37(a)–(g) at the respective times $t = 0.5, 1.0, 1.5, 2.0, 2.5, 3.0, 3.5$. The result is quite different from that in Figures 3.35 and 3.36

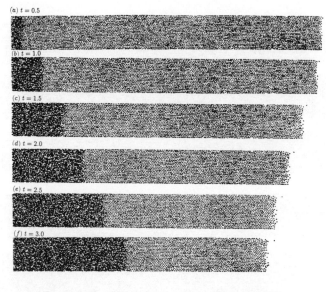

Figure 3.36.   Molecular speed $= -60$, no damping.

Figure 3.37.   Molecular speed $= -60$, parallel damping.

in that areas of rarefaction develop near the upper and lower walls resulting in a rounded shock front in Figure 3.37(b) and jagged shock fronts in Figures 3.37(c)–(g).

Note that Examples 3 and 6 show that, on the molecular level, nonconservative shock simulation yields general results quite different from conservative simulation. Indeed, the very same damping in Examples 2 and 6 yield different results which depend on the relative frame of reference, which is usually not the case in the large.

Chapter 4

N(Number of Molecules) > 10000.
The Cavity Problem

4.1.   Introduction

We assume now that the number of *molecules* N is very large, i.e. $N > 10000$, and assume that various molecular potentials, which will be discussed in the next four chapters, are known. Our approach will be to aggregate the molecules into particles using mass and energy conservation. This approach is what engineers call the *lumped mass* technique. The particles will then be studied dynamically by means of molecular type formulas, that is, formulas which include both attraction and repulsion. By using this methodology all the machinery developed in the last chapter can be used in the present one. However, it will now be essential to allow for gravity.

4.2.   Particle Arrangement and Equations for Water Vapor

Consider first a 30 cm by 240 cm rectangle and on it construct a regular triangular grid with 8479 points, shown in Figure 4.1, by the recursion formulas

$$x(1) = -15.0, \quad y(1) = 0.0$$

$$x(i) = x(i-1) + 1.0, \quad y(i) = 0.0, \qquad\qquad i = 2, 31$$

$$x(32) = -14.5, \quad y(32) = 0.866$$

$$x(i) = x(i-1) + 1.0, \quad y(i) = 0.866, \qquad\quad i = 33, 61$$

$$x(i) = x(i-61), \quad y(i) = y(i-61) + 1.732, \quad i = 62, 8479.$$

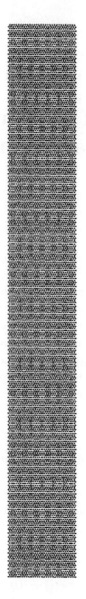

Figure 4.1.   8479 particles.

The side of each triangle in the grid has length $1\,\mathrm{cm}$. At each grid point $(x(i), y(i))$ set a particle P_i, that is an aggregate of molecules. Thus, the distance between any two immediate neighbors is unity. Since the initial particle positions are known, we also assign each particle a small initial velocity of 0.00001 in the X or Y direction, determined at random, with its sign $(\pm)$ also determined at random

The force in dynes between two distinct particles P_i and P_j which are R_{ij} cm apart will be taken to have magnitude F given by

$$F(R_{ij}) = -\frac{G}{R_{ij}^3} + \frac{H}{R_{ij}^5}, \qquad G > 0, \quad H > 0. \tag{4.1}$$

The justification for this choice is as follows. First we wish to choose a formula which is qualitatively like (3.3) so that it too is composed of attraction and repulsion components. The choices of the exponents 3 and 5 guarantees further that the volatile motion between molecules with exponents 7 and 13 will not prevail for the more massive particles.

Thus, from (4.1),

$$\phi(R_{ij}) = -\frac{G}{2R_{ij}^2} + \frac{H}{4R_{ij}^4} \text{ (ergs)}. \tag{4.2}$$

Our first problem is to determine A and B. Assume that $F(1) = 0$, so that, from (4.1),

$$-G + H = 0. \tag{4.3}$$

In order to determine a second equation, some relevant observations must be made first.

Note that the number N of water vapor molecules which can be arranged in the rectangle using a regular triangular grid is

$$N = \frac{30}{(3.06)10^{-8}} \cdot \frac{240}{(2.65)10^{-8}} = (8.87)10^{18}. \tag{4.4}$$

Also, note that since the mass of a water molecule is $(30.103)10^{-24}\,\mathrm{gr}$, the total mass M of the water molecules inside the 30 cm by 240 cm rectangle in Figure 4.1 is

$$M = (2.67)10^{-4}\,\mathrm{gr}. \tag{4.5}$$

Distributing this mass over the 8479 particles for conservation of total mass yields an individual particle mass m of

$$m = (3.15)10^{-8} \text{ gr.} \tag{4.6}$$

From (3.2) and (4.4), the total potential energy E_M of the molecular configuration is, approximately,

$$E_M = 3 \sum_{i=1}^{(8.88)10^{18}} \left\{ (1.9646)10^{-13} \left[\left(\frac{2.725}{3.06} \right)^{12} - \left(\frac{2.725}{3.06} \right)^6 \right] \right\}$$

$$= -(1.3)10^6 \text{ erg.} \tag{4.7}$$

On the other hand, the total potential energy E_p of the particle configuration is, from (4.2), approximately,

$$E_p = 3 \sum_{i=1}^{8479} \left(-\frac{1}{2}G + \frac{1}{2}H \right) = 25437 \left(-\frac{1}{2}G + \frac{1}{2}H \right). \tag{4.8}$$

However, the kinetic energies of both the particle and molecular configurations are relatively negligible so that total energy is conserved by setting $E_M = E_p$. Thus the second equation for G and H is

$$25437 \left(-\frac{1}{2}G + \frac{1}{2}H \right) = -(1.3)10^6. \tag{4.9}$$

The solution of (4.3) and (4.9) is $G = H = 205$. Thus, (4.1) takes the particular form

$$F(R_{ij}) = 205 \left(-\frac{1}{R_{ij}^3} + \frac{1}{R_{ij}^5} \right).$$

We assume next that two particles interact only within the local interaction distance $D = 1.3 \text{ cm}$, which is the solution of the equation $\frac{dF}{dR_{ij}} = 0$.

The dynamical equation of motion for each particle P_i of the system is then given by

$$\frac{d^2 \vec{R}_i}{dt^2} = -980\vec{\delta} + \frac{\alpha}{m} \sum_{\substack{j \\ j \neq 1}} (205) \left[\frac{1}{R_{ij}^5} - \frac{1}{R_{ij}^3} \right] \frac{\vec{R}_{ji}}{R_{ij}}; \quad R_{ij} < D, \tag{4.10}$$

in which $\vec{\delta} = (0, 1), \alpha$ is a parameter, and $i = 1, 8479$. The reason for the introduction of the parameter α is that particle interaction should be local relative to gravity, that is, gravity must dominate for R_{ij} less than, but close to, unity, which we assume to mean, at present, for $R_{ij} \geq 0.7$. However, this is the case by choosing $\alpha = m$, since, for $R_{ij} = 0.9, 0.8, 0.7$, and 0.6, F takes the values 66, 225, 622, and 1687, respectively , the first three of which are less than 980. Thus, the dynamical equations for the particles are

$$\frac{d^2 \vec{R}_i}{dt^2} = -980\vec{\delta} + \sum_{\substack{j \\ j \neq 1}} (205) \left[\frac{1}{R_{ij}^5} - \frac{1}{R_{ij}^3} \right] \frac{\vec{R}_{ji}}{R_{ij}}; \quad R_{ij} < D. \tag{4.11}$$

It should be noted that other choices of α are under study.

4.3. Particle Equilibrium

We now allow the 8479 particles in Figure 4.1 to find their own equilibrium when interacting in accordance with (4.11). We choose $\Delta t = 0.0001$ and use the no slip reflection protocol.

The initial motion of the system is almost one of free fall. So, for the first 20000 time steps, each velocity is damped by the factor 0.2 every 20000 time steps. For the next 20000 time steps, each velocity is damped by the factor 0.4 every 20000 time steps. For the third 20000 time steps, each velocity is damped by the factor 0.7 every 20000 steps. For the final 20000 steps the damping is removed. In this fashion, the particles are eased down into a stable configuration. Finally, to obtain a square set of particles, all particles with $y_i > 30$ are removed to yield the 7549 particle set shown in Figure 4.2. The positions and velocities of these particles are used as the initial data for the cavity flow examples to be described next.

4.4. Examples

In the cavity examples which follow, we use the same protocol as was used for molecules in Section 3.4 with respect to wall reflection and averaging velocities.

Example 1. Consider now the cavity problem for the 7459 particles in Figure 4.2. Let $V = -40$, $\Delta t = 0.00001$ and $J = 20000$. Then Figures 4.3–4.6 show the development of a classical primary vortex at the

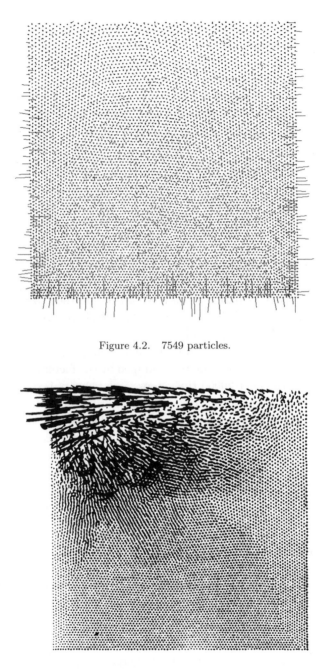

Figure 4.2.   7549 particles.

Figure 4.3. $t = 0.2$.

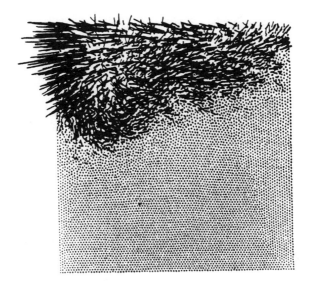

Figure 4.4. $t = 0.6$.

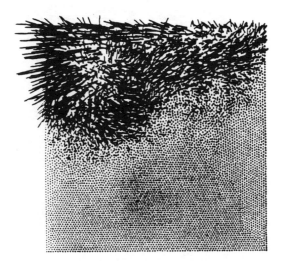

Figure 4.5. $t = 1.0$.

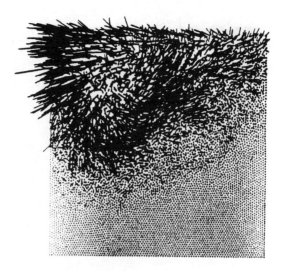

Figure 4.6. $t = 1.4$.

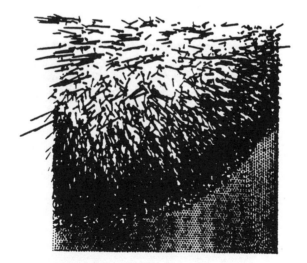

Figure 4.7. $V = -3000$, $t = 0.04$.

Figure 4.8. $V = -3000, \ t = 0.08$.

respective times $t = 0.2, 0.6, 1.0, 1.4$. Entirely similar results followed for $J = 14000, 11000, 9000$.

Results analogous to those just described were obtained also with $V = -100, \ -250, -600$, but the size of the primary vortex increased with the wallspeed.

Example 2. To simulate turbulence, set $V = -3000, \Delta t = 5(10)^{-7}, J = 80000$. Figures 4.7–4.10 show the flow development at the respective times 0.04, 0.08, 0.12, 0.15. Figure 4.7 shows that the motion begins with a compression wave. Figures 4.8 and 4.9 show that the ensuing particle repulsion is up and to the right in the usual primary vortex direction. Figure 4.10 shows the large crosscurrent over the usual primary vortex direction. Again, to support the contention that Figure 4.10 represents fully turbulent flow, we use the concept of a small vortex stated in Section 3.6. In searching for small vortices, attention was confined to within a circle of radius 1 cm around each particle. At $t = 0.15$, there resulted 355 small vortices which are shown in Figure 4.11. Moreover, Figure 4.12 shows the distribution of 349 small vortices at the time $t = 0.18$, so that, after only

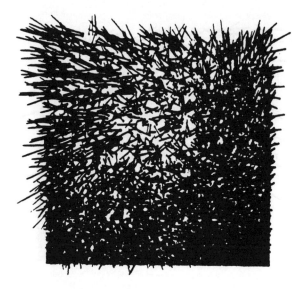

Figure 4.9. $V = -3000$, $t = 0.12$.

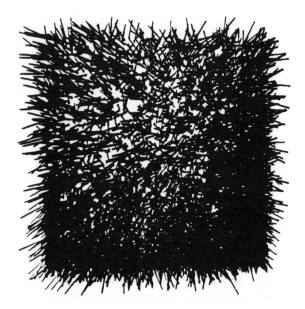

Figure 4.10. $V = -3000$, $t = 0.15$.

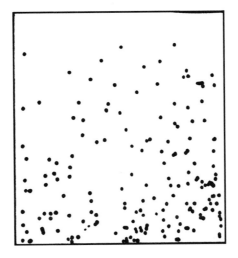

Figure 4.11.   Small vortices (355) at $t = 0.15$.

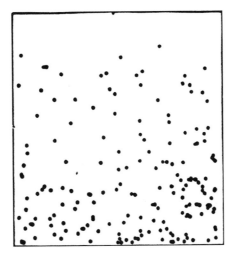

Figure 4.12.   Small vortices (349) at $t = 0.18$.

$0.03\,\mathrm{sec}$, this figure shows the rapid change throughout the cavity of the vortex distribution shown in Figure 4.11.

Results entirely similar to Figures 4.7–4.11 followed with $J = 60000$, 40000, 20000.

Chapter 5

N(Number of Molecules) > 10000.
Crack and Fracture Development

5.1. Introduction

The problem to be discussed in this chapter is development of a crack in a stressed plate. Specifically, we will simulate crack and fracture development in a stressed, slotted copper plate. Note that copper is a soft metal. Note also that, whereas the simulation in Section 3.8 was accomplished with molecules, in this chapter we accomplish our simulation with particles.

5.2. Formula Derivation

An approximate potential function for the interaction of two close copper atoms is

$$\phi(r_{ij}) = \frac{1.55104(10)^{-8}}{r_{ij}^{12}} - \frac{1.398068(10)^{-10}}{r_{ij}^{6}} \text{ erg.} \tag{5.1}$$

From (5.1) it follows that the magnitude F of the force $\vec{F}$, in dynes, between the copper atoms is

$$F(r_{ij}) = \frac{1.861248(10)}{r_{ij}^{13}} - \frac{8.388408(10)^{-2}}{r_{ij}^{7}}. \tag{5.2}$$

The minimum ϕ occurs when $F(r_{ij}) = 0$, and is at $r = 2.46\,\text{Å}$. This yields

$$\phi(2.46) = -3.15045(10)^{-13} \text{ erg.} \tag{5.3}$$

With these observations made, consider a rectangular copper plate which is 8 cm by 11.4 cm. To simulate the plate, let the points P_i, with respective coordinates (x_i, y_i) be given by

$$x(1) = -3.9, \quad y(1) = -5.7158$$
$$x(41) = -4.0, \quad y(41) = -5.5426$$
$$x(i+1) = x(i) + 0.2, \quad y(i+1) = y(1), \qquad\qquad i = 1, 2, \ldots, 39$$
$$x(i+1) = x(i) + 0.2, \quad y(i+1) = y(41), \qquad\qquad i = 41, 42, \ldots, 80$$
$$x(i) = x(i-81), \quad y(i) = y(i-81) + 2(0.1732), \quad i = 82, 83, \ldots, 2713.$$

The resulting arrangement is shown in Figure 5.1. The (x_i, y_i) are vertices of a regular triangular mosaic in which the distance from any P_i to an immediate neighbor is 0.2 cm. The P_i are assumed now to be particles of an 8 cm by 11.43 cm rectangular copper plate. The neighbors of any P_i are defined to be the neighbors of P_i for all time.

To determine a mass m for each P_i, we use total mass conservation. Suppose the plate is filled with copper atoms using, again, a regular triangular mosaic, but one in which the distance between two immediate neighbors is 2.46 Å. Then the approximate number N^* of atoms in the plate is

$$N^* = 1.745(10^{17}). \tag{5.4}$$

Since the mass of a copper atom is $1.0542(10^{-22})$ g, the total mass M of the copper atoms is $M = 1.840(10^{-5})$ g. Distributing this mass over the

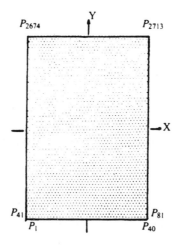

Figure 5.1. The initial configuration. (From: D. Greenspan, Particle Modeling, Birkhauser, Boston, 1997, pp. 165–166.)

particles yields a particle mass m given by

$$m = 6.782(10^{-9})\,\text{g}. \tag{5.5}$$

To determine force and potential formulas, we use energy conservation. Since the minimum potential between two copper atoms is given by (5.3), it follows under the assumption of negligible kinetic energy (as in ideal crystal theory (Megaw (1973))), that the total energy E^* of the system of atoms is, approximately,

$$E^* = -1.6493(10^5)\,\text{erg}. \tag{5.6}$$

Assume now that the force $\vec{F}$, in dynes, between the particles has magnitude F given by

$$F(R_{ij}) = -\frac{G}{R_{ij}^3} + \frac{H}{R_{ij}^5}, \quad G > 0, \ H > 0 \tag{5.7}$$

in which R_{ij} is measured in centimeters. Hence

$$\phi(R_{ij}) = -\frac{G}{2R_{ij}^2} + \frac{H}{4R_{ij}^4} \ (\text{ergs}). \tag{5.8}$$

Assuming $\phi(R_{ij})$ is minimal for $R_{ij} = 0.2$, so that $F(0.2) = 0$, implies

$$-\frac{G}{(0.2)^3} + \frac{H}{(0.2)^5} = 0. \tag{5.9}$$

Approximating the total energy E of the particle system yields

$$E = 3(2713)\left(-\frac{G}{2(0.2)^2} + \frac{H}{4(0.2)^4}\right) \ (\text{ergs}). \tag{5.10}$$

Equating E and E^* implies

$$-\frac{G}{2(0.2)^2} + \frac{H}{4(0.2)^4} = -20.264. \tag{5.11}$$

The solution of (5.9) and (5.11) is $G = 3.24224$, $H = 0.12969$, so the (5.7) takes the specific form

$$F(R_{ij}) = -\frac{3.24224}{R_{ij}^3} + \frac{0.12969}{R_{ij}^5}. \tag{5.12}$$

The force between two particles should be local in the presence of gravity, so we introduce now a normalizing constant α such that at a convenient distance of $0.34\,\text{cm}$, the force between two particles is small relative to gravity. This is essential because we have assumed that forces act only between neighbors, and initially the distance between two neighbors is 0.2 cm. If we define "small relative to gravity" to mean 0.1% of the effect of gravity, and assume that

$$\alpha| - 3.24224/(0.34)^3 + 0.12969/(0.34)^5| < (0.001)980\,m, \qquad (5.13)$$

then α is approximately $(1.25)10^{-10}.$ Since one may consider the plate to be fully supported, the dynamical equation for the motion of each particle is then

$$m\frac{d^2\vec{R}_i}{dt^2} = 1.25(10^{-10}) \sum_{\substack{j \\ j \neq i}} \left[\frac{0.12969}{R_{ij}^5} - \frac{3.24224}{R_{ij}^3} \right] \frac{\vec{R}_{ji}}{R_{ij}}, \qquad (5.14)$$

in which the summation is taken only over the neighbors of P_i. From (5.5) and the introduction of the computationally convenient transformations $R^* = 4R, T^2 = 10t^2$, Eq. (5.14) reduces to

$$\frac{d^2\vec{R}_i^*}{dT^2} = \sum_{\substack{j \\ j \neq i}} \left[\frac{0.97908}{(R_{ij}^*)^5} - \frac{1.52981}{(R_{ij}^*)^3} \right] \frac{\vec{R}_{ji}^*}{R_{ij}^*}. \qquad (5.15)$$

For the distance D^* at which a particle bond breaks, we choose, as recommended by Ashurst and Hoover (1976), the value R^* at which dF/dR^* first becomes negative. From (5.15), then, $D^* = 1.033$.

5.3.   Example

The purpose of the example described now is to ascertain where a plate with a slot in it will crack first. Since many plates used in, say, aerodynamics and structural mechanics, have holes for insertion of screws or bolts, such a simulation informs one of the portions of the plate that need special support. Consider then the slotted copper plate in Figure 5.2 in which the 15 particles P_i, $i = 1070 + 41\,k, k = 0, 1, 2, \ldots, 14$ have been removed. At each time step the particles on the bottom row will be relocated so that their Y coordinates are decreased by 0.00002 units and the particles on the top row will be relocated so that their Y coordinates are increased by 0.00002 units . Thereby, the plate is stretched. This is accomplished readily because copper is a soft metal. Solving (5.15) numerically with $\Delta T = 0.0001$ yields the following results. Figure 5.3 shows the developing force field throughout the bottom half of the plate at the time $T = 6.0$. We need only consider the

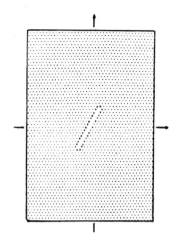

Figure 5.2. The slotted plate. (From: D. Greenspan, Particle Modeling, Birkhauser, Boston, 1997, pp. 165–166.)

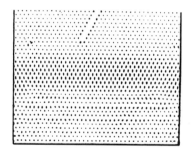

Figure 5.3. $T = 6.0$. (From: D. Greenspan, Particle Modeling, Birkhauser, Boston, 1997, pp. 165–166.)

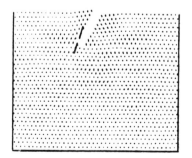

Figure 5.4. $T = 12.8$. (From: D. Greenspan, Particle Modeling, Birkhauser, Boston, 1997, pp. 165–166.)

bottom half of the plate because of the symmetry of the formulation. In our figures force fields are represented by vectors emanating for the centers of the particles. Figure 5.4, at $T = 12.8$, shows clearly from the force field that the first crack occurs at the lower left of the slot, and hence by symmetry, simultaneously at the upper right.

For additional examples and discussion see Greenspan (1997), Chapter 13.

Chapter 6

N(Molecules) > 10000.
Contact Angle of Adhesion

6.1. Introduction

In Chapter 4, we studied the dynamical behavior of a *liquid* by aggregating molecules into particles. In Chapter 5 we did the same for a *solid*. In this chapter we study the dynamical interaction of a *solid* and a *liquid* using a particle formulation. The problem is one of interest in materials science and concerns the angle of contact which a liquid drop makes with a solid surface, the so called contact angle of adhesion. A peculiarity of the present study is that a potential is available for the liquid and a potential is available for the solid, but no potential is available for the interaction of the liquid with the solid. To resolve this problem we will resort to approximate methodology devised by chemists. We consider the problem in three dimensions. The liquid will be water while the solid will be graphite. The discussion follows the work of M. Korlie (1996).

6.2. Local Force Formulas

Figure 6.1 shows the initial model for a horizontal graphite solid. In this five layer solid, the particles are arranged on a regular tetrahedral grid of edge length 0.03834 cm and height 0.031306 cm. The length, width and height of the solid are 1.22688 cm, 1.1952 cm and 0.125224 cm, respectively. There are 5949 graphite particles and 0.0 is the Z coordinate of each point in the bottom layer. The Z axis is perpendicular to the graphite surface.

Figure 6.2 shows the initial model for a water drop. The particles in this drop are arranged on a regular tetrahedral grid with edge length 0.0611742 cm and height 0.0499486 cm. There are 1265 water particles in this relatively spherical drop. The horizontal radius of the drop contains

Figure 6.1.   The initial graphite solid.

Figure 6.2.   The initial water drop.

6 particles, so that the radius of the drop is approximately

$$6(0.611742) = 0.3670452 \, \text{cm}. \tag{6.1}$$

The center of the drop is at $(0,0,0)$ and the particles are arranged symmetrically with respect to the Y axis.

Let the force $\boldsymbol{F_g}$, in dynes, between two graphite particles R cm apart have magnitude F_g given by

$$F_g(R) = -\frac{G}{R^3} + \frac{H}{R^5}, \tag{6.2}$$

where G, H are positive constants to be determined. Let $\phi_g(R)$ be a potential for two graphite particles R cm apart. Then,

$$\phi_g(R) = -\frac{G}{2R^2} + \frac{H}{4R^4}. \tag{6.3}$$

Assuming that $F_g = 0$ between two immediate neighbors for the graphite particles in Figure 6.1 then implies that

$$-(0.03834)^2 G + H = 0. \tag{6.4}$$

Assuming negligible kinetic energy for the graphite system yields an approximate total system energy E_g given by

$$E_g = 6(5949)\left(-\frac{G}{2(0.03834)^2} + \frac{H}{4(0.03834)^4}\right). \tag{6.5}$$

To determine unique values for G, H, we use (6.4), (6.5) and conservation of total energy as follows. For two graphite atoms which are r Å apart (Girafalco and Lad (1991), Kelly (1981))

$$\phi(R) = \left(-\frac{24.3}{r^6} + \frac{38591.3}{r^{12}}\right) 10^{-12} \text{ erg}. \tag{6.6}$$

Hence, the magnitude $F(r)$ of the local force between two graphite atoms which are r Å apart is given by

$$F(r) = \left(-\frac{145.8}{r^7} + \frac{463095.6}{r^{13}}\right) 10^{-4} \text{ dynes}. \tag{6.7}$$

Note that $F(3.834) = 0$.

We next fill the solid with graphite atoms which are placed at the vertices of a regular tetrahedral grid with edge length 3.834 Å and height 3.130448 Å. The height of a regular triangle forming the base of each regular tetrahedral grid is 3.230341 Å. Thus, the number N of atoms which fill the surface is, approximately,

$$N = \frac{1.22688}{(3.834)10^{-8}} \frac{1.1952}{(3.320341)10^{-8}} \frac{0.125224}{(3.130448)10^{-8}},$$

or

$$N = (4.60775)10^{21}.$$

Assuming negligible kinetic energy for the atomic system, the total energy E of the atomic system is, approximately,

$$E = 6(4.60775)10^{21}\left(-\frac{24.3}{(3.834)^6} + \frac{38591.3}{(3.834)^{12}}\right)10^{-12} \text{ erg,}$$

or,

$$E = -(1.0575558)10^8 \text{ erg.} \tag{6.8}$$

Equating E and E_g yields two linear equations for G, H, the solution of which is $G = 17.420968$, $H = (2.560805)10^{-2}$. Thus

$$F_g = -\frac{17.420968}{R^3} + \frac{2.560805}{R^5}10^{-2}, \tag{6.9}$$

$$\phi_g(R) = -\frac{8.710484}{R^2} + \frac{6.4020125}{R^4}10^{-3}. \tag{6.10}$$

Note that (6.10) can be written as

$$\phi_g(R) = 4\epsilon_g\left[-\left(\frac{\sigma_g}{R}\right)^2 + \left(\frac{\sigma_g}{R}\right)^4\right], \tag{6.11}$$

where $\sigma_g = (2.711047)10^{-2}$, $\epsilon_g = (2.96284)10^3$.

The mass of a graphite atom is approximately $(1.9938)10^{-23}$ gr, so that the total atomic mass, when distributed over the 5949 graphite particles yields a graphite particle mass M_g given by

$$M_g = (1.544282)10^{-5} \text{ gr.} \tag{6.12}$$

Since $F_g(0.03834) = 0$, the equilibrium distance for the graphite particles is

$$R_g = 0.03834 \text{ cm.} \tag{6.13}$$

Let the force $\boldsymbol{F_w}$, in dynes, between two water particles R cm apart have magnitude F_w given by

$$F_w(R) = -\frac{G}{R^3} + \frac{H}{R^5}, \tag{6.14}$$

where G, H are positive constants to be determined. Let $\phi_w(R)$ be a potential for two water particles R cm apart. Then,

$$\phi_w(R) = -\frac{G}{2R^2} + \frac{H}{4R^4}. \tag{6.15}$$

Assuming that $F_g = 0$ between two immediate neighbors for the water particles in Figure 6.2 then implies that

$$-(0.0611742)^2 G + H = 0. \tag{6.16}$$

Assuming negligible kinetic energy for the water system yields an approximate total system energy E_w given by

$$E_w = 6(1265) \left(-\frac{G}{2(0.0611742)^2} + \frac{H}{4(0.0611742)^4} \right). \tag{6.17}$$

For actual water molecules which are r Å apart we use the approximation

$$\phi(R) = (1.9646383) \left(-\frac{2.725^6}{r^6} + \frac{2.725^{12}}{r^{12}} \right) 10^{-13} \text{ erg.} \tag{6.18}$$

Hence, the magnitude $F(r)$ of the local force between two water molecules which are r Å apart is given by

$$F(r) = (1.9646383) \left(-6\frac{2.725^6}{r^7} + 12\frac{2.725^{12}}{r^{13}} \right) 10^{-5} \text{ dynes.} \tag{6.19}$$

Note that $F(3.05871) = 0$.

We next fill the sphere with water molecules which are placed at the vertices of a regular tetrahedral grid with edge length 3.05871 Å. Thus, the number N of atoms which fill the region is, approximately,

$$N = \frac{4\pi}{3} \left(\frac{0.3670452}{(3.05871)10^{-8}} \right)^3,$$

or,

$$N = (7.238229)10^{21}.$$

Assuming negligible kinetic energy for the molecular system, the total energy E of the system is, approximately,

$$E = 6(7.2382229)10^{21}(1.9646383)10^{-13}$$
$$\times \left(-\left(\frac{2.725}{3.05871} \right)^6 + \left(\frac{2.725}{3.05871} \right)^{12} \right) \text{ erg,}$$

or,

$$E = -(2.133075)10^9 \text{ erg.} \tag{6.20}$$

Equating E and E_w yields two linear equations for G, H, the solution of which is $G = 4206.887899, H = 15.743364$. Thus

$$F_w(R) = -\frac{4206.887899}{R^3} + \frac{15.743364}{R^5}, \tag{6.21}$$

$$\phi_w(R) = -\frac{2103.44395}{R^2} + \frac{3.935841}{R^4}. \tag{6.22}$$

Note that (6.10) can be written as

$$\phi_w(R) = 4\epsilon_w \left[-\left(\frac{\sigma_w}{R}\right)^2 + \left(\frac{\sigma_w}{R}\right)^4 \right], \tag{6.23}$$

where $\sigma_w = (4.325669)10^{-2}, \epsilon_w = (2.810376)10^5$.

The mass of a water molecule is approximately $(30.103)10^{-24}$ gr, so that the total molecular mass, when distributed over the 1265 water particles, yields a water particle mass M_w given by

$$M_w = (1.722469)10^{-4} \text{ gr.} \tag{6.24}$$

Since $F_w(0.0611742) = 0$, the equilibrium distance for the water particles is

$$R_w = 0.0611742 \text{ cm.}$$

To determine the local force $\boldsymbol{F_{gw}}$ between a water particle and a graphite particle R cm apart, we use the empirical law of bonding (Hirschfelder, Curtiss and Bird (1967)):

$$\phi_{gw}(R) = 4\epsilon_{gw} \left[-\left(\frac{\sigma_{gw}}{R}\right)^2 + \left(\frac{\sigma_{gw}}{R}\right)^4 \right], \tag{6.25}$$

where $\epsilon_{gw} = (\epsilon_g \epsilon_w)^{\frac{1}{2}}$ and $\sigma_{gw} = \frac{1}{2}(\sigma_g + \sigma_w)$. Thus, $\epsilon_{gw} = (2.885601)10^4$, $\sigma_{gw} = (3.518358)10^{-2}$, and

$$\phi_{gw}(R) = -\frac{142.8816071}{R^2} + \frac{0.1768709}{R^4}.$$

The magnitude of $\boldsymbol{F_{gw}}$ is

$$F_{gw} = -\frac{285.7632142}{R^3} + \frac{0.7074836}{R^5},$$

with $F_{gw}(0.0497571) = 0$. Hence, $R_{gw} = 0.0497571$ cm is the equilibrium distance for water particle and graphite particle interaction.

6.3.   Dynamical Equations

We assume that each particle P_i is acted upon locally only by those particles within a sphere of radius two equilibrium distances and centered at P_i. This radius is called the distance of local interaction.

The motion of a graphite particle P_i as it interacts with other graphite particles is given by

$$M_g \frac{d^2 \boldsymbol{R}_i}{dt^2} = -980 M_g \boldsymbol{\delta} + \alpha_g \sum_j \left[\left(-\frac{17.420968}{R_{ij}^3} + \frac{2.560805}{R_{ij}^5} 10^{-2} \right) \frac{\boldsymbol{R}_{ji}}{R_{ij}} \right]$$

$$(6.26)$$

where the summation is taken over particles P_j, different from P_i, which are within the prescribed distance D_g (distance of local interaction for the graphite particles) from P_i, R_{ij} is the distance between P_i and P_j, α_g is a scaling factor which assures that the particle interactions are local, that is, small relative to gravity, M_g is the mass of a graphite particle, and $\boldsymbol{\delta} = (0, 0, 1)$. Division by M_g yields from (6.26)

$$\frac{d^2 \boldsymbol{R}_i}{dt^2} = -(980)\boldsymbol{\delta} + \alpha_g \sum_j \left[\left(-\frac{1.128095}{R_{ij}^3} 10^6 + \frac{1.65825}{R_{ij}^5} 10^3 \right) \frac{\boldsymbol{R}_{ji}}{R_{ij}} \right].$$

$$(6.27)$$

Assume that any particle P_j located at more than two equilibrium distances away from a given particle P_i has, relative to gravity, a negligible effect on P_i. Hence, $D_g = 2(0.03834) = 0.07668\,\text{cm}$. By "local relative to gravity" we mean

$$\alpha_g \left| -\frac{1.128095}{(0.07668)^3} 10^6 + \frac{1.65825}{(0.07668)^5} 10^3 \right| = 5\% \,(980). \qquad (6.28)$$

From (6.28) it follows that $\alpha_g = (2.6111721)10^{-8}$, so that (6.27) reduces to

$$\frac{d^2 \boldsymbol{R}_i}{dt^2} = -(980)\boldsymbol{\delta} + \sum_j \left[\left(-\frac{2.94565}{(R_{ij})^3} 10^{-2} + \frac{4.329976}{(R_{ij})^5} 10^{-5} \right) \frac{\boldsymbol{R}_{ji}}{R_{ij}} \right].$$

$$(6.29)$$

For computational convenience, we now make the change of variables

$$\boldsymbol{R}_i = 10 \boldsymbol{R}_i^*, \quad T = 10t.$$

Thus, finally, the dynamical equation for the interaction of two graphite particles is

$$\frac{d^2\boldsymbol{R}_i^*}{dT^2} = -(98.0)\boldsymbol{\delta}$$

$$+ \sum_j \left[\left(-\frac{2.94565}{(R_{ij}^*)^3} + \frac{0.4329976}{(R_{ij}^*)^5} \right) \frac{\boldsymbol{R}_{ji}^*}{R_{ji}^*} \right], \quad i = 1, 2, \ldots, 5949$$

(6.30)

and the distance of interaction is now $D^* = 0.7668\,\text{cm}$.

Using the same reasoning described above, the dynamical equation for the interaction of two water particles becomes

$$\frac{d^2\boldsymbol{R}_i^*}{dT^2} = -(98.0)\boldsymbol{\delta}$$

$$+ \sum_j \left[\left(-\frac{11.96547}{(R_{ij}^*)^3} + \frac{4.4778167}{(R_{ij}^*)^5} \right) \frac{\boldsymbol{R}_{ji}^*}{R_{ji}^*} \right], \quad i = 1, 2, \ldots, 1265.$$

(6.31)

The equilibrium distance is $R_w^* = 0.611742$ and the distance of local interaction is $D_w^* = 1.223484\,\text{cm}$.

Again, using the above reasoning, the dynamical equation for the interaction of a water particle and a graphite particle, where the graphite particle is held fixed, becomes

$$\frac{d^2\boldsymbol{R}_i^*}{dT^2} = -(98.0)\boldsymbol{\delta}$$

$$+ \sum_j \left[\left(-\frac{6.438578}{(R_{ij}^*)^3} + \frac{1.594043}{(R_{ij}^*)^5} \right) \frac{\boldsymbol{R}_{ji}^*}{R_{ji}^*} \right], \quad i = 1, 2, \ldots, 1265,$$

(6.32)

the equilibrium distance is $R_{gw} = 0.497571$, and the distance of local interaction is $D_{gw} = 0.995142\,\text{cm}$.

6.4.   Drop and Slab Stabilization

The graphite surface in Figure 6.1 is stabilized in accordance with (6.30) using the leap frog formulas. Each graphite particle is assigned a zero initial velocity. The time step used is $\Delta T = 5(10)^{-5}$. In order to maintain the solid state, during the numerical procedure we damp all velocities by

Figure 6.3. The stabilized graphite solid.

the factor 0.25 whenever the total system kinetic energy exceeds 100. The graphite system is allowed to run to T_{25000}. At this time the system has contracted to the physically stable configuration shown in Figure 6.3. The height of the configuration is approximately seven-tenths of that shown in Figure 6.1.

To stabilize the water drop, we need a reflection method that keeps the particles within the sphere during the drop stabilization using (631). We accomplish this as follows. Let R be the radius of the relatively spherical water drop in Figure 6.2. At time T_k, let

$$R_{i,k} = \left(X_{i,k}^2 + Y_{i,k}^2 + Z_{i,k}^2 \right)^{\frac{1}{2}},$$

where $(X_{i,k}, Y_{i,k}, Z_{i,k})$ is the position of P_i at time T_k. If $R_{i,k} > R$, the position and velocity of P_i are reset as follows: $D_{i,k} = 2R - R_{i,k}$, $X_{i,k} \rightarrow (D_{i,k})(X_{i,k})/R_{i,k}$, $V_{i,k,x} \rightarrow -0.9 V_{i,k,x}$, $Y_{i,k} \rightarrow (D_{i,k})(Y_{i,k})/R_{i,k}$, $V_{i,k,y} \rightarrow -0.9 V_{i,k,y}$, $Z_{i,k} \rightarrow (D_{i,k})(Z_{i,k})/R_{i,k}$, $V_{i,k,z} \rightarrow -0.9 V_{i,k,z}$, where $(V_{i,k,x}, V_{i,k,y} V_{i,k,z})$ is the velocity of P_i at T_k.

Each of the water particles is assigned zero initial velocity. Set $\Delta T = 5(10)^{-5}$ and allow the water particles to interact in accordance with (6.31) using the leapfrog formulas. The system of water particles exhibits large contraction and expansion modes during the stabilization process and to overcome this all velocities are damped by the factor 0.9 every 200 time steps. At the end of 20000 time steps the damping is removed and the particles are allowed to interact 26000 more time steps, at which time the oscillation modes are no longer present. The resulting physically stable water drop is shown in Figure 6.4. In this figure, the outermost particles have a lower density than the inner particles, which is characteristic of liquid surface tension. The average diameter is approximately seven-tenths of that in Figure 6.2.

Figure 6.4.   The stabilized water drop.

6.5.   Sessile Drop Formation

We next translate the water drop in Figure 6.4 so that it sits centrally on the graphite surface in Figure 6.3. This is done by raising the water drop 3.9 and is shown in Figure 6.5. The graphite particles are now held fixed while the water particles are allowed to interact with themselves and with the graphite particles. In applying the leap frog formulas for the numerical solution of the differential equations, we use $\Delta T = 1(10)^{-6}$ for the first 200,000 time steps and thereafter use $\Delta T = 2(10)^{-6}$. There is an increase in energy due to gravity and to counter this increase all velocities are damped 0.9 every 2000 time steps. This is not considered to be significant because we are interested only in a relatively steady state configuration. The results are given in Figures 6.6–6.9 at the times $T = 0.3, 0.5, 0.7, 0.834$. The figures show the settling of the drop on the graphite surface. There is no seepage of water into the graphite solid.

From Figure 6.9, using a linear least square approximation in the X^*Z^* plane with the lower four boundary particles, we get Figure 6.10. Because of the symmetry with respect to the Y^* axis, the left contact angle and the right are the same. This angle is calculated to be $60.01°$. A reported (Adamson (1976)) experimentally determined contact angle is $60°$.

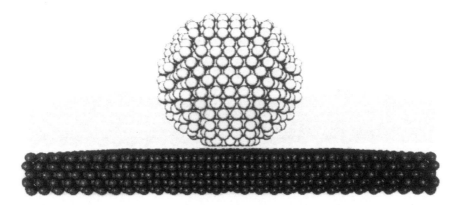

Figure 6.5.   The water drop on the graphite surface.

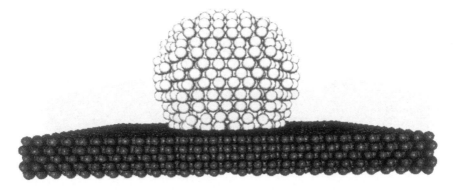

Figure 6.6. $T = 0.3$.

Figure 6.7. $T = 0.5$.

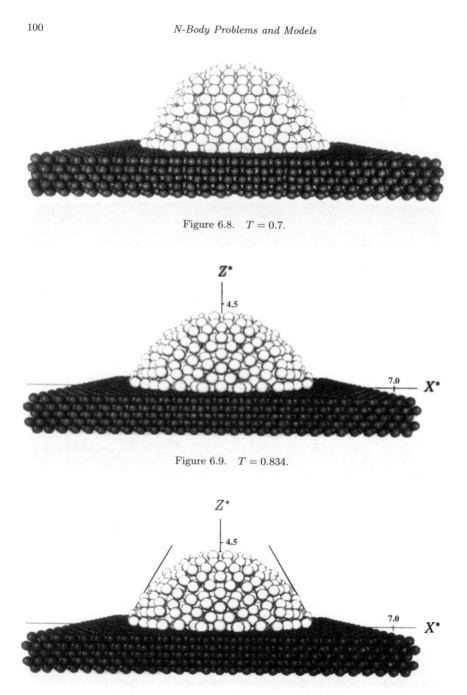

Figure 6.8. $T = 0.7$.

Figure 6.9. $T = 0.834$.

Figure 6.10.   Determination of the contact angle.

6.6. Remark

In addition to sessile drops, pendant, or, hanging, drops are also of interest. For a particle model of a pendant drop and its fall from a position of equilibrium see Greenspan (1997), Chapter 9.

Chapter 7

A Particle Model of Carbon Dioxide Bubbles in Water

7.1. Introduction

In this chapter we will study the emergence of a gas from the interior of a fluid. The gas will be carbon dioxide and the fluid will be water. For simplicity, the simulation will be two dimensional, but all the ideas and methods extend directly to three dimensions. However, the resulting amount of computation would then require greater computing power than to be used at present.

7.2. Formulation of the Particle Model

For two water molecules P_i, P_j, which are r_{ij} Å apart, we again assume the potential and force given by (3.2) and (3.3), that is

$$\phi(r_{ij}) = (1.9646)10^{-13} \left[\frac{2.725^{12}}{r_{ij}^{12}} - \frac{2.725^6}{r_{ij}^6} \right] \text{erg} \quad \left(\frac{\text{grcm}^2}{\text{sec}^2} \right). \quad (7.1)$$

The force $\vec{F}_{ij}$ exerted on P_i by P_j is

$$\vec{F}_{ij} = (1.9646)10^{-5} \left[\frac{12(2.725^{12})}{r_{ij}^{13}} - \frac{6(2.725^6)}{r_{ij}^7} \right] \frac{\vec{r}_{ji}}{r_{ij}} \text{dynes} \quad \left(\frac{\text{grcm}}{\text{sec}^2} \right).$$
$$(7.2)$$

The magnitude F_{ij} of (7.2) is

$$F_{ij} = (1.9646)10^{-5} \left[\frac{12(2.725^{12})}{r_{ij}^{13}} - \frac{6(2.725^6)}{r_{ij}^7} \right] \quad (7.3)$$

and the equilibrium distance is

$$r = 3.059\,\text{Å}. \tag{7.4}$$

We now consider a two dimensional rectangular basin with base 36.708 cm and height 31.791 cm, as shown in Figure 7.1, where we have superimposed an xy coordinate system on the basin. The choice of width and height will simplify later calculations. The basin lies in the upper half of the plane and is symmetrical about the y axis.

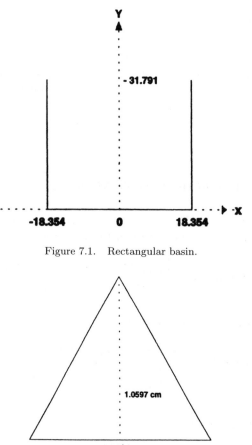

Figure 7.1.   Rectangular basin.

Figure 7.2.   Regular triangular building block.

Using the regular triangular building block shown in Figure 7.2, we now construct a regular triangular grid with edge length 1.2236 cm, as shown in Figure 7.3. Again, this choice will simplify later calculations. The grid has 31 rows and 946 grid points, which are numbered left to right on each row and bottom to top. At each grid point we place a water particle. We now fill the basin with water molecules, but use (7.4) as the edge length of the regular triangular grid and water molecules are placed at the vertices of the resulting grid. The building block for this configuration is shown in Figure 7.4. Hence the number of water molecules in the basin is approximately

$$N = \frac{(36.708)(31.791)}{(3.059)10^{-8}(2.649)10^{-8}} = (1.4401)10^{18}. \tag{7.5}$$

Since the mass of a water molecule is $(30.103)10^{-24}$ g, distributing the total molecular mass over the particles yields a particle mass M_w given by

$$M_w = (4.5826)10^{-8} \text{ g}. \tag{7.6}$$

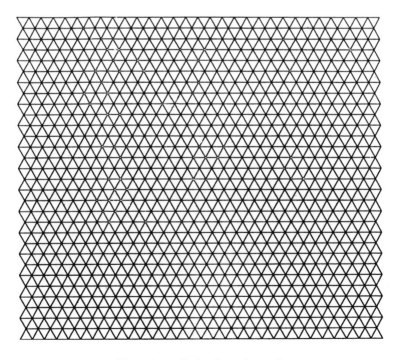

Figure 7.3.   Basin triangular grid.

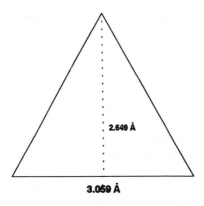

2.649 Å

3.059 Å

Figure 7.4.   Regular triangular building block.

We next assume that the magnitude F of the local force, in dynes, between two water particles P_i, P_j, which are R_{ij} cm apart, is given by

$$F = \frac{G}{R_{ij}} + \frac{H}{R_{ij}^3}.$$ (7.7)

From (7.7), the corresponding potential is

$$\phi(R_{ij}) = -G\ln(R_{ij}) + \frac{H}{2R_{ij}^2}.$$ (7.8)

Now, assuming that $F = 0$ at $R_{ij} = 1.2236\,\text{cm}$, Eq. (7.7) yields

$$(1.2236)^2 G + H = 0.$$ (7.9)

Assuming negligible kinetic energy, the potential energy E for the molecular system is approximately

$$E = 3 \sum_{1}^{(1.4401)10^{18}} \left\{ (1.9646)10^{-13} \left[\left(\frac{2.725}{3.059}\right)^{12} - \left(\frac{2.725}{3.059}\right)^{6} \right] \right\},$$

or,

$$E = -(2.1219)10^5 \,\text{erg}.$$ (7.10)

We now assign each particle a small initial random velocity of either $\pm 10^{-7}$ in the x direction. Thus the kinetic energy is relatively negligible and the energy of the particle system is approximately

$$E = 3 \sum_{1}^{946} \left[-G \ln(1.2236) + \frac{H}{2(1.2236)^2} \right]$$

or,

$$E = -572.7G + 947.771H. \tag{7.11}$$

Equations (7.10) and (7.11) yield

$$-572.7G + 947.771H = -(2.1219)10^5. \tag{7.12}$$

The solution of (7.9) and (7.12) is $G = 106.537, H = -159.507$.

Thus (7.7) now becomes

$$F = \frac{106.537}{R_{ij}} - \frac{159.507}{R_{ij}^3}. \tag{7.13}$$

Using (7.6) and (7.13), let the motion of a water particle P_i as it interacts with other water particles be determined by the dynamical equation

$$M_w \frac{d^2 \vec{R}_i}{dt^2} = -980 M_w \vec{\delta} + \alpha \sum_{\substack{j \\ j \neq i}}^{946} \left[\frac{106.537}{R_{ij}} - \frac{159.507}{R_{ij}^3} \right] \frac{\vec{R}_{ji}}{R_{ij}},$$

where $\vec{\delta} = (0, 1)$, and α is a normalization constant. Thus

$$\frac{d^2 \vec{R}_i}{dt^2} = -980 \vec{\delta} + \alpha \sum_{\substack{j \\ j \neq i}}^{946} \left[\frac{2.32482}{R_{ij}} 10^9 - \frac{3.48071}{R_{ij}^3} 10^9 \right] \frac{\vec{R}_{ji}}{R_{ij}}. \tag{7.14}$$

We choose α this time so that each P_i in the very top row of the configuration is supported completely by local interaction with any particle that lies $1.05967\,\mathrm{cm}$ directly below it. Thus,

$$\alpha \left[\frac{2.32482}{1.05967} 10^9 - \frac{3.48071}{(1.05967)^3} 10^9 \right] = 980,$$

which yields

$$\alpha = -(1.34)10^{-6}. \tag{7.15}$$

Hence, substitution of (7.15) into (7.14) yields

$$\frac{d^2 \vec{R}_i}{dt^2} = -980 \vec{\delta} + \sum_{\substack{j \\ j \neq i}}^{946} \left[-\frac{3115.26}{R_{ij}} + \frac{4664.15}{R_{ij}^3} \right] \frac{\vec{R}_{ji}}{R_{ij}}. \tag{7.16}$$

Making the change of variables

$$T = 10t, \tag{7.17}$$

Eq. (7.16) becomes

$$\frac{d^2 \vec{R}_i}{dT^2} = -9.8\vec{\delta} + \sum_{\substack{j \\ j \neq i}}^{946} \left[-\frac{31.1526}{R_{ij}} + \frac{46.6415}{R_{ij}^3} \right] \frac{\vec{R}_{ji}}{R_{ij}}. \tag{7.18}$$

Equation (7.18) is the one to be used in actual computation.

7.3. Basin of Water Stabilization

We let the basin of water particles seek its own physical stabilization in accordance with (7.18) and the leap frog formulas as follows. Fix $\Delta T = 2(10^{-4})$. Every 500 time steps each velocity is damped by the factor 0.1 and particles are reflected due to wall collision symmetrically with velocity damped by the factor 0.1. The results are shown in Figures 7.5–7.7, where the stable basin of water is shown in Figure 7.7 at $T = 20$.

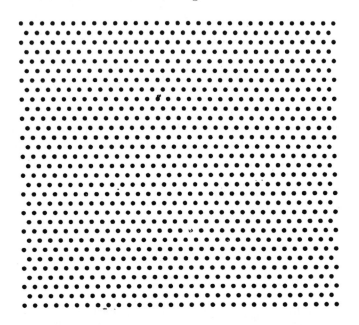

Figure 7.5. H_2O basin at $T = 0$.

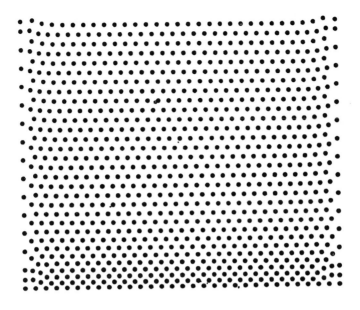

Figure 7.6. H_2O basin at $T = 4$.

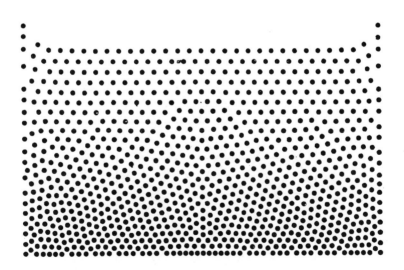

Figure 7.7. H_2O basin at $T = 20$.

7.4.   Carbon Dioxide Bubbles and Their Motions

We now construct a model for carbon dioxide particles and derive dynamical equations for their motions.

In three dimensions at $0°C$, CO_2 gas has a density approximately 500^{-1} that of water (Sears and Zemansky (1957)). Thus, in two dimensions, the density of CO_2 gas will be approximately $500^{-2/3}$, or, approximately $1/63$ that of water. Therefore, from (7.5),

$$(1/63)N = (1/63)(1.4401)^{18} = (2.28587)10^{16}$$

CO_2 molecules are placed in the basin. We next let the number of CO_2 particles in the basin be 33, where these 33 particles are placed at the vertices of a regular triangular grid with edge length 7.3416 cm and height 6.358 cm. This regular triangular grid is shown in Figure 7.8. With this choice we have sufficient CO_2 particles to describe the rise of the bubbles and the motion of the water near the bubbles. The choice of edge length and height of the triangular building block is based on the 33 CO_2 particles for the purpose of placing them evenly in the basin.

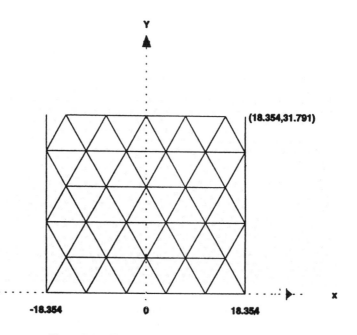

Figure 7.8.   Triangular grid for CO_2 particles.

The potential for two CO_2 molecules, P_i, P_j, which are r_{ij} Å apart is given (Hirschfelder, Curtiss and Bird (1967)) by

$$\phi(r_{ij}) = (1.132051)10^{-13} \left[\frac{4.07^{12}}{r_{ij}^{12}} - \frac{4.07^6}{r_{ij}^6} \right]. \tag{7.19}$$

The magnitude of the force $\vec{F}_{ij}$ exerted on P_i by P_j is then

$$F_{ij} = (1.132051)10^{-5} \left[\frac{(12(4.07)^{12})}{r_{ij}^{13}} - \frac{(6(4.07)^6)}{r_{ij}^7} \right]. \tag{7.20}$$

From (7.20), $F_{ij}(4.568) = 0$. Hence, 4.568 Å is the equilibrium distance. Using these results, an approximate total molecular potential energy of the CO_2 molecular system is given by

$$E = 3 \sum_1^{(2.28587)10^{16}} \left\{ (1.132051)10^{-13} \left[\left(\frac{4.07}{4.568} \right)^{12} - \left(\frac{4.07}{4.568} \right)^6 \right] \right\},$$

or,

$$E = -1940.79 \, \text{erg}. \tag{7.21}$$

We assume that the magnitude of the local force between two CO_2 particles P_i, P_j, R_{ij} cm apart is given by

$$F = \frac{G}{R_{ij}} + \frac{H}{R_{ij}^3}. \tag{7.22}$$

Thus, from (7.22), the corresponding potential function is

$$\phi(R_{ij}) = -G \ln(R_{ij}) + \frac{H}{2R_{ij}^2}. \tag{7.23}$$

Assuming that $R_{ij} = 7.3416$ cm is the CO_2 particles equilibrium distance, (7.22) yields

$$(7.3416)^2 G + H = 0. \tag{7.24}$$

The potential energy for the CO_2 particle system in the basin is approximately

$$E = 3 \sum_1^{33} \left[-G \ln(7.3416) + \frac{H}{2(7.3416)^2} \right]$$

or,

$$E = -197.362G + 0.91838H. \tag{7.25}$$

Equating (7.21) and (7.25), we find

$$-197.362G + 0.91838H = -1940.79. \tag{7.26}$$

The solution of (7.24) and (7.26) is $G = 7.86185, H = -423.746$. Thus, (7.22) becomes

$$F = \frac{7.86185}{R_{ij}} - \frac{423.746}{R_{ij}^3}. \tag{7.27}$$

The mass of a CO_2 molecule is $(7.3585)10^{-23}$ g. Thus the mass of a CO_2 particle is

$$M_c = \frac{(2.28587)10^{16}(7.3585)10^{-23}}{33},$$

or,

$$M_c = (5.097)10^{-8} \text{ g}. \tag{7.28}$$

The dynamical equation of a CO_2 particle is given by

$$M_c \frac{d^2 \vec{R}_i}{dt^2} = -980 M_c \vec{\delta} + \alpha \sum_{\substack{j \\ j \neq i}}^{33} \left[\frac{7.86185}{R_{ij}} - \frac{423.746}{R_{ij}^3} \right] \frac{\vec{R}_{ji}}{R_{ij}}. \tag{7.29}$$

From (7.15) and (7.28), Eq. (7.29) yields

$$\frac{d^2 \vec{R}_i}{dt^2} = -980 \vec{\delta} + \sum_{\substack{j \\ j \neq i}}^{33} \left[-\frac{206.688}{R_{ij}} + \frac{11140.3}{R_{ij}^3} \right] \frac{\vec{R}_{ji}}{R_{ij}}. \tag{7.30}$$

By (7.17) and (7.30) we find

$$\frac{d^2 \vec{R}_i}{dT^2} = -9.8 \vec{\delta} + \sum_{\substack{j \\ j \neq i}}^{33} \left[-\frac{2.06688}{R_{ij}} + \frac{111.403}{R_{ij}^3} \right] \frac{\vec{R}_{ji}}{R_{ij}}, \tag{7.31}$$

which is the equation for dynamical interaction between two CO_2 particles.

For CO_2–H_2O interaction we use a simplistic averaging procedure, to be described shortly. We also impose a local interaction distance $D = 1.234$ cm to force local interaction only. This choice is based on the edge length in Figure 7.2 for the purpose of assuring one equilibrium distance of local interaction for the water particles.

From the above discussion we now summarize the dynamical approach to be used. Let P_i, P_j be any two particles in the stable basin shown in Figure 7.7. The motion of P_i is determined by the dynamical equation

$$\frac{d^2\vec{R}_i}{dT^2} = -9.8\vec{\delta} + \sum_{\substack{j \\ j \neq i}}^{946} \left[-\frac{A}{R_{ij}} + \frac{B}{R_{ij}^3} \right] \frac{\vec{R}_{ji}}{R_{ij}}. \tag{7.32}$$

If $R_{ij} > 1.234\,\text{cm}$, then $A = B = 0$. If $R_{ij} < 1.234\,\text{cm}$, then we determine A and B as follows. If the two particles are water particles, then $A = 31.1526, B = 46.6415$. If the two particles are carbon dioxide particles, then $A = 2.06688, B = 111.403$. In all other cases, because the empirical bonding law is not always valid (Hirschfelder, Curtiss and Bird (1967)), we use the following simplistic averaging procedure:

$$A = \frac{1}{2}(31.1526 + 2.06688) = 13.6097,$$

$$B = \frac{1}{2}(46.6415 + 111.403) = 79.0223.$$

For bubble simulation we proceed as follows. Consider the stable basin of H_2O particles in Figure 7.7. We assume that 33 of the 946 particles are now CO_2 particles. No changes in position or velocities are made. The initial configuration is shown in Figure 7.9. The system (7.32) is then solved numerically with $\Delta T = 2(10)^{-4}$ by the leap frog formulas through $T = 340$. The result which shows the emergence of the bubbles is shown in Figures 7.10–7.14 at the indicated times.

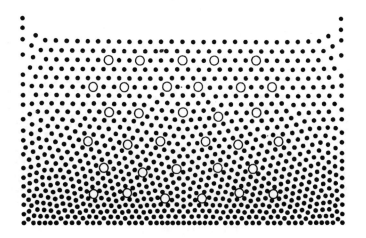

Figure 7.9.   Initial CO_2–H_2O configuration.

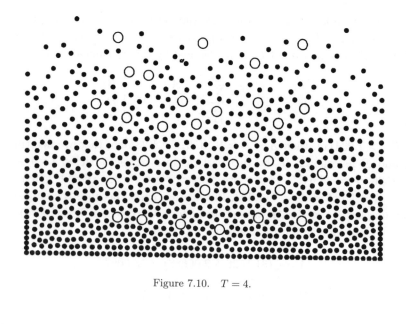

Figure 7.10. $T = 4$.

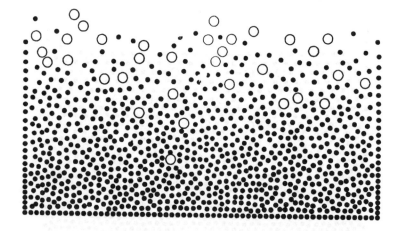

Figure 7.11. $T = 20$.

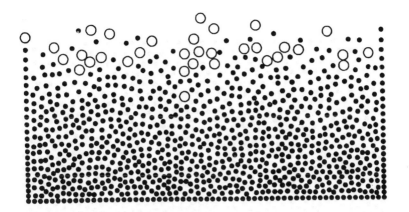

Figure 7.12. $T = 40$.

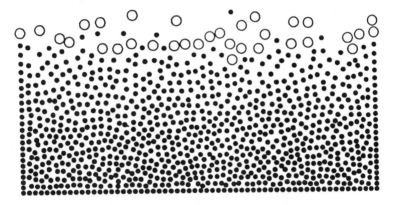

Figure 7.13. $T = 120$.

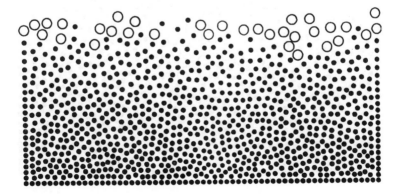

Figure 7.14. $T = 340$.

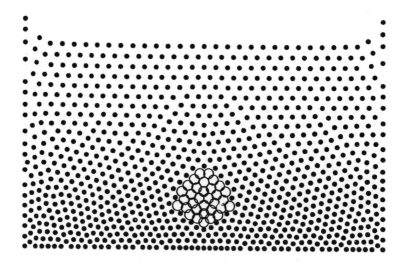

Figure 7.15.   Initial CO_2–H_2O configuration.

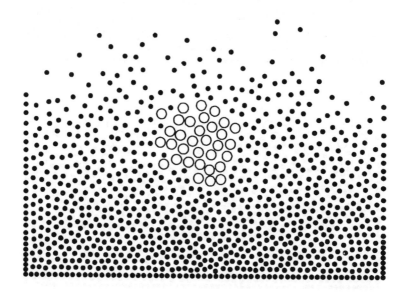

Figure 7.16. $T = 4$.

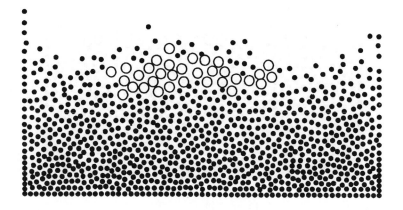

Figure 7.17. $T = 6$.

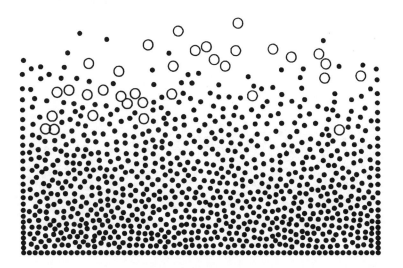

Figure 7.18. $T = 12$.

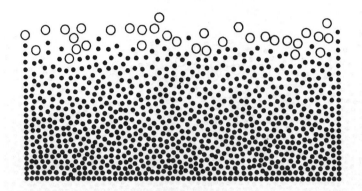

Figure 7.19. $T = 240$.

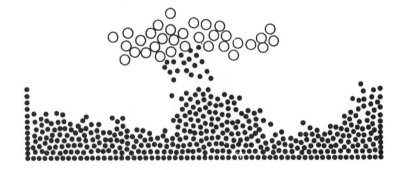

Figure 7.20. Wake flow at $T = 6$.

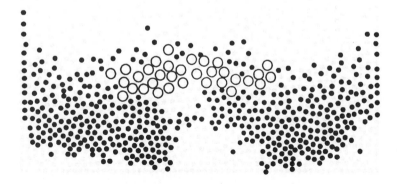

Figure 7.21. Vertical flow of uppermost particles at $T = 6$.

Finally, the 33 CO_2 particles are set in the position shown in Figure 7.15, where the effect is to create a large, compressed CO_2 bubble. For $\Delta T = 2(10)^{-4}$ the resulting motion is shown in Figures 7.16–7.21 at the indicated times. Figure 7.16 shows the immediate compression wave effect directly above the bubble at the basin surface. The figures show the disintegration of the bubble as it rises. Figure 7.20 shows at $T = 6$ only those H_2O particles that were originally not above the bubble and their formation into a wake below the rising CO_2 bubble. Similarly, Figure 7.21 shows at $T = 6$ those H_2O particles originally above the CO_2 bubble and how they have moved downward toward the area vacated by the particles in the wake. A large rotational H_2O motion is evident at this time.

Chapter 8

A Particle Model a Dodecahedral Rotating Top

8.1. Introduction

Rigid body motion is of fundamental interest in mathematics, science, and engineering. In this chapter we will introduce a new, simplistic approach to this area of study. We will consider a discrete dodecahedral body and simulate its motion when it spins like a top. The approach will not require the use of special coordinates, Cayley–Klein parameters, tensors, dyadics, or related concepts (Goldstine (1980)). All that will be required is Newtonian mechanics in three dimensional Euclidean XYZ space. As is desirable in simulating a top's motion, the numerical methodology will conserve exactly the same energy, linear momentum and angular momentum as does the associated differential system.

8.2. A Discrete, Dodecahedral Top

Consider, in XYZ space as shown in Figure 8.11, a dodecahedron $P_1 - P_8$. For the present, let the distances between the vertices be given as follows:

$$\|P_1 P_j\| = \|P_j P_8\| = 2 \, \text{cm}, \quad j = 2 - 7$$
$$\|P_2 P_3\| = \|P_3 P_4\| = \|P_4 P_5\| = \|P_5 P_6\| = \|P_6 P_7\| = \|P_7 P_8\| = 1 \, \text{cm}$$
$$\|P_2 P_5\| = \|P_3 P_6\| = \|P_4 P_7\| = 2 \, \text{cm}.$$

The vertices $P_2 - P_7$ are taken to be vertices of a regular, plane hexagon with edge length $1 \, \text{cm}$. The hexagon is located in the plane $z = \sqrt{3}$ and the vertex coordinates are $P_2(0, 1, \sqrt{3})$, $P_3(\frac{1}{2}\sqrt{3}, \frac{1}{2}, \sqrt{3})$, $P_4(\frac{1}{2}\sqrt{3}, -\frac{1}{2}, \sqrt{3})$, $P_5(0, -1, \sqrt{3})$, $P_6(-\frac{1}{2}\sqrt{3}, -\frac{1}{2}, \sqrt{3})$, $P_7(-\frac{1}{2}\sqrt{3}, \frac{1}{2}, \sqrt{3})$. The geometric center

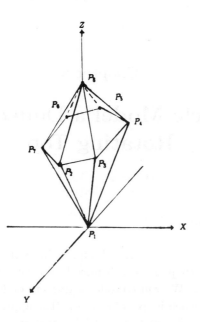

Figure 8.1.   The dodecahedron.

of the hexagon is the point $Q(0, 0, \sqrt{3})$. The point P_8, located on the line through $P_1 Q$, is at $P_8(0, 0, 2\sqrt{3})$.

Next, we define the neighbors of $P_1 - P_8$. The neighbors of P_1 are defined to be $P_2 - P_7$. The neighbors of P_8 are defined to be $P_2 - P_7$. The neighbors of any one of the points $P_2 - P_7$ are taken to be P_1, P_8, the two adjacent points in the hexagon and the point opposite it in the hexagon. Thus, for example, as seen in Figures 8.1 and 8.2, the neighbors of P_2 are P_1, P_8, P_3, P_7 and P_5. The reason for this choice of neighbors for each point in the set $P_2 - P_7$ is that it is sufficient to preserve rigidity during the extensive computations to be described.

In order to set the top in rotation, the initial velocities of $P_2 - P_7$ are chosen to be, as shown in Figure 8.2, respectively, $\vec{v}_2 = (v, 0, 0)$, $\vec{v}_3 = (\frac{1}{2}v, -\frac{1}{2}v\sqrt{3}, 0)$, $\vec{v}_4 = (-\frac{1}{2}v, -\frac{1}{2}v\sqrt{3}, 0)$, $\vec{v}_5 = (-v, 0, 0)$, $\vec{v}_6 = (-\frac{1}{2}v, \frac{1}{2}v\sqrt{3}, 0)$, $\vec{v}_7 = (\frac{1}{2}v, \frac{1}{2}v\sqrt{3}, 0)$, so that the velocity of each particle is perpendicular to the line joining it to Q. The points P_1 and P_8 are given zero initial velocities.

Next, we want to tilt the top and this is done by rotating the XZ plane through an angle α, as shown in Figure 8.3. Thus, the initial positions (x'_i, y'_i, z'_i) and initial velocities $(v_{i,x'}, v_{i,y'}, v_{i,z'})$ of $P_1 - P_8$ are determined

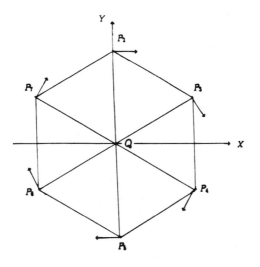

Figure 8.2.   The planar hexagon and initial velocities.

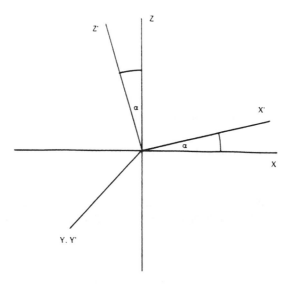

Figure 8.3.   The rotation of the XZ plane.

finally by

$$x'_i = x_i \cos \alpha + z_i \sin \alpha, \quad y'_i = y_i, \quad z'_i = -x_i \sin \alpha + z_i \cos \alpha,$$

$$v_{ix'} = v_{ix} \cos \alpha + v_{iz} \sin \alpha, \quad v_{iy'} = v_{iy}, \quad v_{iz'} = -v_{ix} \sin \alpha + v_{iz} \cos \alpha.$$

Thus, once the parameters v and α are given, all initial data for a tilted, rotating dodecahedral top are determined.

8.3.   Dynamical Equations

The motion of our rotating top is now treated as a eight-body problem. At any time t, let $P_i, i = 1 - 8$, be located at $\vec{r}_i = (x_i, y_i, z_i)$, have velocity $\vec{v}_i = (\dot{x}_i, \dot{y}_i, \dot{z}_i) = (v_{ix}, v_{iy}, v_{iz})$, and have acceleration $\vec{a}_i = (\ddot{x}_i, \ddot{y}_i, \ddot{z}_i) = (\dot{v}_{ix}, \dot{v}_{iy}, \dot{v}_{iz})$. For $i \neq j$, let $\vec{r}_{ij}$ be the vector from P_i to P_j and let r_{ij} be the magnitude of $\vec{r}_{ij}, i = 1 - 8; j = 1 - 8; i \neq j$. Let $\phi = \phi(r_{ij})$ be a potential function defined by the pair $P_i, P_j, i \neq j$. Then, for $i = 1 - 8$, the Newtonian dynamical equations for the motion of the particles $P_1 - P_8$ are the second order differential equations [1]

$$m_i \vec{a}_i = \sum_{nbrs} \left(-\frac{\partial \phi}{\partial r_{ij}} \frac{\vec{r}_i - \vec{r}_j}{r_{ij}} \right) - 980 m_i \vec{\delta}, \quad i = 1 - 8, \qquad (8.1)$$

in which $\vec{\delta} = (0, 0, 1)$ for $i = 2 - 8, \vec{\delta} = (0, 0, 0)$ for $i = 1$, and the summation is taken only over neighbors. Note that P_1 is fixed to move only in the $Z = 0$ plane.

Equations (8.1) are fully conservative, that is, they conserve system energy, linear momentum and angular momentum.

8.4.   Numerical Method

In order to solve system (8.1) in a fashion which conserves the same energy, linear momentum and angular momentum, as in Section 2.2, we first rewrite (8.1) as the following equivalent first order differential system:

$$\frac{d\vec{r}_i}{dt} = \vec{v}_i$$

$$m_i \frac{d\vec{v}_i}{dt} = \sum_{nbrs} \left(-\frac{\partial \phi}{\partial r_{ij}} \frac{\vec{r}_i - \vec{r}_j}{r_{ij}} \right) - 980 m_i \vec{\delta}.$$

Now fix a time step Δt. Set $t_k = k \Delta t, k = 0, 1, 2, 3, \ldots$. At t_k let P_i be at $\vec{r}_{i,k} = (x_{i,k}, y_{i,k}, z_{i,k})$ with velocity $\vec{v}_{i,k} = (v_{i,x,k}, v_{i,y,k}, v_{i,z,k})$. Then the differential system will be approximated by

$$\frac{\vec{r}_{i,k+1} - \vec{r}_{i,k}}{\triangle t} = \frac{\vec{v}_{i,k+1} + \vec{v}_{i,k}}{2}$$

$$m_i \frac{\vec{v}_{i,k+1} - \vec{v}_{i,k}}{\triangle t} = \sum_{nbrs} \left[\left(-\frac{\phi(r_{ij,k+1}) - \phi(r_{ij,k})}{r_{ij,k+1} - r_{ij,k}} \right) \right.$$

$$\left. \times \left(\frac{\vec{r}_{i,k+1} + \vec{r}_{i,k} - \vec{r}_{j,k+1} - \vec{r}_{j,k}}{r_{ij,k+1} + r_{ij,k}} \right) \right] - 980 m_i \vec{\delta}.$$

These difference equations, at each time step, consist of 48 nonlinear equations in the unknowns $x_{i,k+1}, y_{i,k+1}, z_{i,k+1}, v_{i,x,k+1}, v_{i,y,k+1}, v_{i,z,k+1}$, $i = 1 - 8$; and in the knowns $x_{i,k}, y_{i,k}, z_{i,k}, v_{i,x,k}, v_{i,y,k}, v_{i,z,k}$. These are solved readily by Newton's method.

Moreover, since we will assure the rigidity of the motion, the trajectories of the points P_1 and Q will fully characterize the motion of the top, thus providing the simple graphical procedures which will be used. The coordinates of the point Q will be taken to be $(x*, y*, z*)$. We assume that the motion has terminated whenever $z* < 0.866$.

8.5. Examples

In considering examples, we must choose a force function $\vec{F}$ which acts on two neighboring particles P_i, P_j. If the two particles are initially 2 cm apart, we choose the force to have magnitude F_1 given by

$$F_1 = A \left[-\frac{1}{r_{ij}^3} + \frac{4}{r_{ij}^5} \right], \quad A > 0, \tag{8.2}$$

in which A is sufficiently large insure that the attraction and repulsion inherent in (8.2) occur over very short periods of time. This is also necessary to assure rigidity and the value of A will be discussed later when a numerical time step will be chosen. Notice also that $F_1(2) = 0$.

If the two particles are initially 1 cm apart, we choose the force to have magnitude F_2 given by

$$F_2 = A \left[-\frac{1}{r_{ij}^3} + \frac{1}{r_{ij}^5} \right]. \tag{8.3}$$

Note that $F_2(1) = 0$.

Thus, the pairs (P_1, P_j), $j = 2 - 7$, will be governed by (8.2); the pairs $(P_2, P_5), (P_3, P_6), (P_4, P_7)$, and $(P_8, P_j), j = 2-7$, will be governed by (8.2); while the pairs $(P_2, P_3), (P_3, P_4), (P_4, P_5), (P_5, P_6), (P_6, P_7), (P_7, P_2)$ will be governed by (8.3).

However, the numerical equations require potentials so that for F_1, F_2 we now choose the corresponding potentials

$$\phi_1 = A \left[-\frac{1}{2r_{ij}^2} + \frac{1}{r_{ij}^4} \right]$$

$$\phi_2 = A \left[-\frac{1}{2r_{ij}^2} + \frac{1}{4r_{ij}^4} \right].$$

In all the examples to be described next, $\Delta t = 10^{-7}$ and $A = 10^9$. These values were determined from computer results which yielded rigid motion. Unless otherwise stated, let $m_i = 1, i = 1 - 8$.

Example 1. Let $\alpha = 15°, v = 400$. The conservative numerical formulas were then run for 200,000,000 time steps with a printout every 100,000 steps. The resulting motion of P_1 for the first cycle is shown in Figure 8.4,

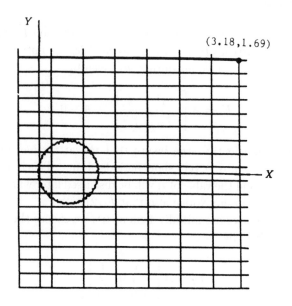

Figure 8.4.   First cycle, $\alpha = 15°, v = 400$.

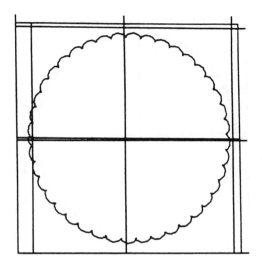

Figure 8.5.　Enlargement of Figure 8.4.

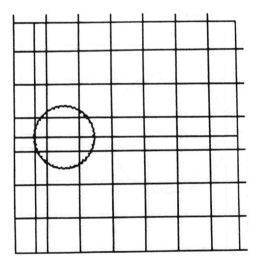

Figure 8.6.　Full trajectory, $\alpha = 15°, v = 400$.

which contains 1086 points, on the region

$$-0.32 \leq x \leq 3.1, \quad -1.71 \leq y \leq 1.71, \tag{8.4}$$

and in the enlarged version in Figure 8.5 on the region

$$-0.196 \leq x \leq 0.916, \quad -0.4678 \leq y \leq 0.4678.$$

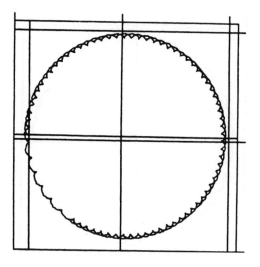

Figure 8.7.   Enlargement of Figure 8.6.

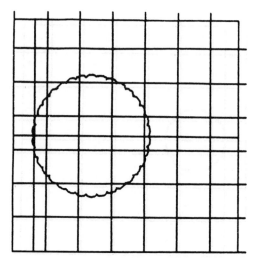

Figure 8.8.   First cycle, $\alpha = 30°, v = 400$.

The motion is characterized by a circular, cycloid type trajectory around the central point $(0.4483, 0)$ in the XY plane. The point (x^*, y^*, z^*) is always on the line $(0.4483, 0, z)$ with z^* varying in the range $1.668 < z^* < 1.673$. The full 2000 point trajectory is shown in Figure 8.6 and in enlarged form in Figure 8.7. Figure 8.7 shows that the motion is not periodic.

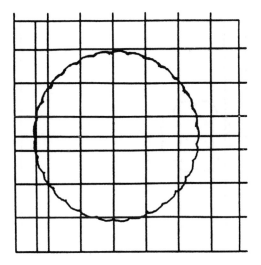

Figure 8.9.   First cycle, $\alpha = 45°, v = 400$.

Example 2.   Let $\alpha = 30°, v = 400$. The conservative numerical formulas were run for 200,000,000 time steps with a printout every 100,000 steps. The resulting motion of P_1 for the first cycle is shown in Figure 8.8 on the region defined by (8.4). The resulting circular, cycloid type motion is centered at (0.8660,0). The point (x^*, y^*, z^*) is always on the line $(0.8660, 0, z)$, with z^* varying in the range $1.4801 < z^* < 1.50$. Thus, the motion is around a larger circle than that in Example 1 and z^* shows greater variation than in Example 1.

Example 3.   Let $\alpha = 45°, v = 400$. The conservative numerical formulas were run for 200,000,000 time steps with a printout every 100,000 steps. Figure 8.9 shows the first cycle of the motion on the region defined by (8.4). The resulting circular, cycloid type motion is centered at (1.2247, 0). The point (x^*, y^*, z^*) is always on the line $(1.2247, 0, z)$, with z^* varying in the range $1.1857 < z^* < 1.2245$. Thus the motion is around a larger circle than that in Example 2 and z^* shows greater variation than in Example 2.

Examples 1–3 show that for fixed $v = 400$, as α increases the radii of the associated circles increase while the cusps become flatter.

Example 4.   Let $\alpha = 60°, v = 400$. The motion terminates almost immediately. Moreover, to the nearest unit, $60°$ is the first value for which the motion terminates, since for $59°$ the top does not fall.

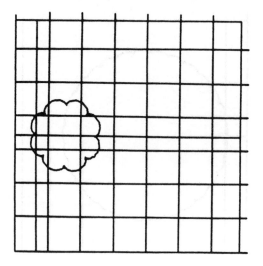

Figure 8.10.   First cycle, $\alpha = 15°, v = 200$.

Example 5.   Let $\alpha = 15°, v = 200$. The conservative numerical formulas were then run for 200,000,000 time steps with a printout every 100,000 steps. Figure 8.10 shows the first cycle of the motion on the region defined by (8.4). The resulting circular, cycloid type motion is centered at (0.4483, 0). The point (x^*, y^*, z^*) is always on the line $(0.4483, 0, z)$, with z^* varying in the range $1.6438 < z^* < 1.6730$, which is larger than that in Example 1. Also, the number of cusps are fewer and larger than those in Example 1, while the number of trajectory points is now only 745.

In the next example, we explore an exceptionally difficult area in the theory of tops, namely, the area of nonhomogeneous tops.

Example 6.   Let $\alpha = 15°, v = 400$, and reset m_2 and m_4 so that $m_2 = 2$, $m_4 = 4$. The conservative numerical formulas were then run for 100,000 time steps with a printout every 20 steps. Figure 8.11 shows the trajectory of P_1, for the first 4000 points, in the region

$$-0.015 \leq x \leq 0.015, \quad -0.02 \leq y \leq 0.02.$$

Figure 8.12 shows an enlarged version of the motion of the first 1000 points in the region

$$-0.008 \leq x \leq 0.1, \quad -0.02 \leq y \leq 0.015.$$

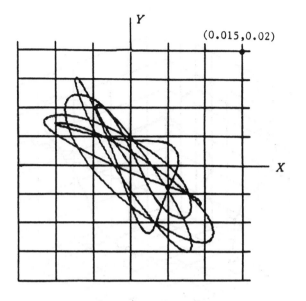

Figure 8.11. $\alpha = 15°, v = 400, m_2 = 2, m_4 = 4$.

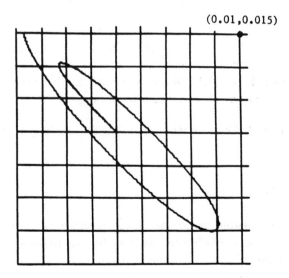

Figure 8.12.   Enlargement of first 1000 point trajectory of Figure 8.11.

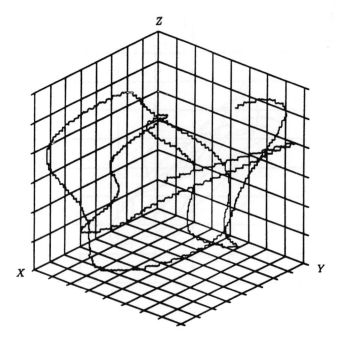

Figure 8.13.   5000 point trajectory for the motion of Q.

Both figures reveal large looping motions for P_1. The graphics procedure available was insufficient to determine whether or not small cusps were part of the trajectory. Figure 8.13 shows the erratic motion over the full 5000 points of Q in the narrow three dimensional range

$$0.444 \leq x^* \leq 0.452, \quad -0.008 \leq y^* \leq 0.008, \quad 1.670 \leq z^* \leq 1.676.$$

8.6.   Remarks

The number of possible variations of the parameters in Section 8.2 is unlimited. The choices which were made enable one to design a least complex computer program for the numerical procedure described, since only two force magnitudes F_1 and F_2 were required. If, for example, one were to choose

$$\|P_1P_2\| = \|P_1P_3\| = \|P_1P_4\| = \|P_1P_5\| = \|P_1P_6\| = \|P_1P_7\| = R\,\mathrm{cm},$$

with $R \neq 1, R \neq 2$, then one would require, in addition, an F_3 with

$$F_3 = A \left[-\frac{1}{r_{ij}^3} + \frac{R^2}{r_{ij}^5} \right], \quad A > 0$$

and an appropriate potential ϕ_3.

The number of possible variations of the α, v, and m_i parameters in Section 8.4 is also unlimited. For those additional choices which we studied, the examples in Section 8.4 showed results typical of the full set of calculations.

Chapter 9

A Particle Model of Self Reorganization

9.1. Introduction

Steinberg (1975) describes several interesting biological experiments in morphogenesis, that is, in the self reorganization of biological cells. For example, Holtfreter showed that embryonic tissue, consisting of distinct endoderm, mesoderm, and ectoderm layers, when separated out, could recombine into tissue with normal endoderm, mesoderm, and ectoderm layers. (See Figure 9.1) As another example, in an experiment by Wilson, cells and cell clusters obtained by squeezing a sponge through a fine silk cloth could reunite and aggregates could reconstruct themselves into functional sponges.

In this chapter we will concentrate on a computer simulation of the Holtfreter experiment.

9.2. The Computer Algorithm

Consider N particles, $P_i, i = 1, 2, 3, \ldots, N$. For $\Delta t > 0$, let $t_k = k\Delta t, k = 0, 1, 2, 3, \ldots$. For each of $i = 1, 2, 3, \ldots, N$, let m_i denote an adhesive measure associated with P_i, and, in two dimensions, let P_i at t_k be located at $\vec{r}_{i,k} = (x_{i,k}, y_{i,k})$, have velocity $\vec{v}_{i,k} = (v_{i,k,x}, v_{i,k,y})$, and have acceleration $\vec{a}_{i,k} = (a_{i,k,x}, a_{i,k,y})$. Let position, velocity and acceleration be related by the leap frog formulas.

At t_k, let the force acting on P_i be $\vec{F}_{i,k}$. We relate force and acceleration by

$$\vec{F}_{i,k} = m_i \vec{a}_{i,k}.$$

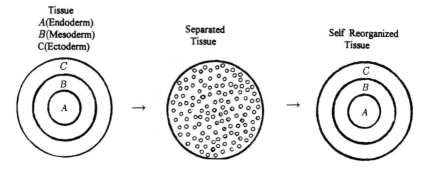

Figure 9.1.   The Holtfreter experiment.

The force $\vec{F}_{ij,k}$ on P_i exerted by P_j at time t_k is assumed to be

$$\vec{F}_{ij,k} = \left(\left(-\frac{G_{ij}}{(r_{ij,k})^p} + \frac{H_{ij}}{(r_{ij,k})^q} \right) \frac{\vec{r}_{ji,k}}{r_{ij,k}} \right),$$

in which the terms G_{ij}, H_{ij} will be allowed to vary. The total force $\vec{F}_{i,k}$ on P_i due to all other particles different from P_i is defined by

$$\vec{F}_{i,k} = \sum_{\substack{j=1 \\ j \neq i}}^{N} \left(\left(-\frac{G_{ij}}{(r_{ij,k})^p} + \frac{H_{ij}}{(r_{ij,k})^q} \right) \frac{\vec{r}_{ji,k}}{r_{ij,k}} \right).$$

Note finally that the introduction of an additional parameter D is essential to assure that particle interactions are local. This is necessary because only local forces were present in the Holtfreter experiment. We will require that whenever $r_{ij,k} > D$, then

$$\vec{F}_{i,k} = \vec{0}(r_{ij,k} > D).$$

9.3.   Example

Since, in general, particles do not adhere when in a gaseous state and are rigid when in a solid state, self reorganization can occur only in a liquid or near-liquid state. Relative to this observation, previous calculations (Greenspan(1980)) allow us now to fix the parameters as follows: $\Delta t = 0.0001$, $p = 3, q = 5$, $G_{ij} = H_{ij} = 5m_i m_j$, $D = 2.2$. For, then, if P_i is to be a liquid particle, the speed v_i of P_i has been deduced for various adhesive measures m_i. In particular;

$$m_i = 2000 \quad \text{implies} \quad 100 \leq v_i \leq 170 \qquad (9.1)$$

$$m_i = 4000 \quad \text{implies} \quad 90 \leq v_i \leq 160 \qquad (9.2)$$

$$m_i = 10000 \quad \text{implies} \quad 50 \leq v_i \leq 80. \qquad (9.3)$$

Consider now a square region in the XY plane whose vertices are $(-16, -16)$, $(-16, 16)$, $(16, 16)$, $(16, -16)$. On this region construct a triangular grid of 1072 points using the recursion formulas

$$
\begin{aligned}
x(1) &= -15.5, \quad y(1) = -16.0 \\
x(i+1) &= x(i) + 1.0, \quad y(i+1) = -16.0, \qquad i = 1, 31 \\
x(33) &= -16.0, \quad y(33) = -15.0 \\
x(i+1) &= x(i) + 1.0, \quad y(i+1) = -15.0, \qquad i = 33, 64 \\
x(i) &= x(i-65), \quad y(i) = y(i-65) + 2.0, \quad i = 66, 1072.
\end{aligned}
$$

This point set is shown in Figure 9.2.

We now fix a set A which consists of 38 particles each with adhesive measure 10000, a set B of 266 particles each with adhesive measure 4000,

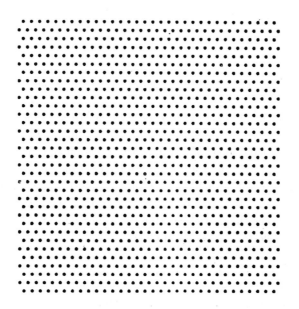

Figure 9.2.　1072 points in a 32 cm by 32 cm square.

Figure 9.3.   The initial data.

and a set C of 768 particles each with adhesive measure 2000. The particles are distributed at the 1072 points shown in Figure 9.2, with no two
particles at the same point. A particle at the point $(x(i), y(i))$ is denoted
by P_i. In Figure 9.3, the A particles, which have the largest adhesive
measures, are denoted by circles; the particles of set B, which have the
intermediate adhesive measures, are denoted by quadrilaterals; and the
particles of set C which have the smallest adhesive measures are denoted
by triangles.

Next a velocity is assigned to each particle. In agreement with (9.1)–
(9.3), each A particle is assigned a speed of 60 while each of the B and C
particles is assigned a speed of 150. The XY direction and the corresponding
$(\pm)$ signs of the velocity vectors are determined at random, and the resulting
velocity is shown in Figure 9.3 as a vector emanating from each particle's
center. For a complete listing of all the initial data, see Greenspan (1988).

In order to keep the particles within the square while they are in motion,
the following reflection rules are applied:

(a) if $x_i > 16$, set $x_i = 32.0 - x_i$, $v_{x,i} = -0.99v_{x,i}$, $v_{y,i} = 0.99v_{y,i}$
(b) if $x_i < -16$, set $x_i = -32.0 - x_i$, $v_{x,i} = -0.99v_{x,i}$, $v_{y,i} = 0.99v_{y,i}$
(c) if $y_i > 16$, set $y_i = 32.0 - y_i$, $v_{x,i} = 0.99v_{x,i}$, $v_{y,i} = -0.99v_{y,i}$
(d) if $y_i < -16$, set $y_i = -32.0 - y_i$, $v_{x,i} = 0.99v_{x,i}$, $v_{y,i} = -0.99v_{y,i}$

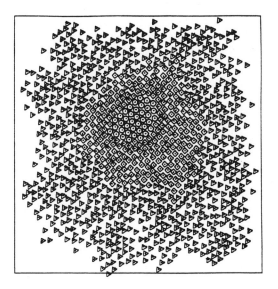

Figure 9.4.   Completed self reorganization. (From: D. Greenspan, Particle Modeling, Birkhauser, Boston, 1997, p. 69.)

The small velocity damping in rules (a)–(d) insures numerical stability when using the time step $\Delta t = 0.0001$.

The resulting self reorganization occurs in the following way. First the A particles organize into small units which then converge to form the central core in Figure 9.4. Then the B particles form a layer around the A particles as shown in Figure 9.4. Finally, after an extended period, the C particles form a third layer around the B particles to complete the self reorganization shown in Figure 9.4. For the extensive details and for additional examples, see Greenspan (1997).

Chapter 10

Particle Model of a Bouncing Elastic Ball

10.1. Introduction

The mathematical problems of contact mechanics are of such difficulty that, as yet, there exists no comprehensive underlying theory for this widely studied area of engineering and physics (Johnson (1985), Klarbring (1988), Lebon and Raous (1992), Raous Chabrand and Lebon (1988), Martins and Oden (1987), Shillor (1988)). Section 3.8 has already examined a contact mechanics problem for fluid microdrops. In this chapter we will examine such a problem for an elastic solid.

10.2. Generating a Ball as a Particle Model

Relative to the origin in the XY plane, consider three circles C_1, C_2, C_3 with respective radii r_1, r_2, r_3 cm. Assume $r_1 < r_2 < r_3$. On C_1 we wish to generate 120 points $(x_i, y_i), i = 1, 120$; on C_2 120 more points $(x_i, y_i), i = 121, 240$; and on C_3 120 more points $(x_i, y_i), i = 241, 360$. On each circle, any two adjacent points should have the same distance apart as any other two adjacent points. Also, we wish the points of C_2 to be staggered between the points of C_1 and C_3. These constraints are satisfied readily by applying, in radian measure, the recursion formulas

$$
\begin{aligned}
x_{i+1} &= r_1 \cos(0.05235987755i) & i &= 0, 119 \\
y_{i+1} &= r_1 \sin(0.05235987755i) & i &= 0, 119 \\
x_{i+1} &= r_2 \cos(0.05235987755i + 0.02617993878) & i &= 120, 239 \\
y_{i+1} &= r_2 \sin(0.05235987755i + 0.02617993878) & i &= 120, 239 \\
x_{i+1} &= r_3 \cos(0.05235987755i) & i &= 240, 359 \\
y_{i+1} &= r_3 \sin(0.05235987755i) & i &= 240, 359.
\end{aligned}
$$

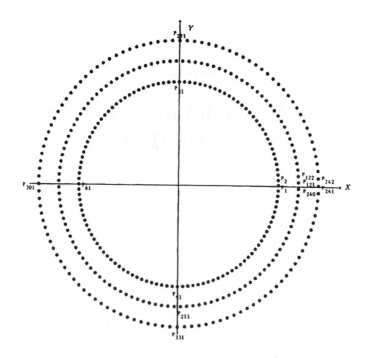

Figure 10.1.   The ordering of the points.

For $r_1 = 5, r_2 = 6, r_3 = 7$, Figure 10.1 shows the resulting 360 point configuration and indicates the counterclockwise ordering of the points.

Next, at each point (x_i, y_i) we set a unit mass particle P_i, as shown in Figure 10.2. This configuration is called a ball. If one wishes the ball to have a more continuous representation, one need only increase the radius of each particle, and this effect is shown in Figure 10.3. In the plot of Figure 10.3, the radius of each particle is taken to be 2.5 times the radius of each particle in Figure 10.2. Both types of representations will be of value in later discussions.

Now we define the neighbors of any particle P_i as follows. If P_i is on the circle C_1, then it will have exactly four neighbors, namely, the two particles adjacent to it on C_1 and the two particles closest to it on C_2. If P_i is on the circle C_2 it will have exactly six neighbors, namely, the two particles adjacent to it on C_2, the two particles closest to it on C_1, and the two particles closest to it on C_3. Finally, if P_i is on the circle C_3, then it will have exactly four neighbors, namely, the two particles adjacent to

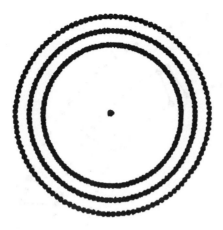

Figure 10.2.   Initial particle configuration (small radius).

Figure 10.3.   Initial particle configuration (large radius).

it on C_3 and the two particles closest to it on C_2. Figure 10.4 shows the neighbors of P_i for typical particles on C_1, C_2, and C_3.

In order to simulate a bouncing ball, we next introduce force interactions between particles. It is assumed throughout that each particle interacts only with its neighbors. In order not to be overly general in the discussion we assume that $r_1 = 5, r_2 = 6, r_3 = 7$. Any other choice can be treated in the fashion to be described.

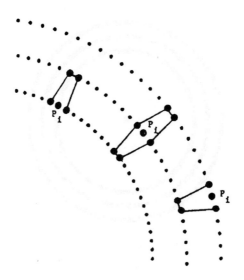

Figure 10.4.   The neighbors of P_i.

First, let P_i be on C_1, let the distance between P_i and any neighbor, say P_j, on C_1, be R_{ij}. The force $\vec{F}_i$ which P_j exerts on P_i is taken to have magnitude F_i given by

$$F_i = -\frac{G}{R_{ij}^3} + \frac{H}{R_{ij}^5}. \tag{10.1}$$

Next note that the distance between P_i and any neighbor on C_1 is $0.2618\,\text{cm}$, which we take to be the equilibrium distance, that is

$$F_i(0.2618) = 0. \tag{10.2}$$

Thus,

$$-\frac{G}{(0.2618)^3} + \frac{H}{(0.2618)^5} = 0$$

which implies $H = (0.2618)^2 G$. If for the moment we choose $G = 1$, then (10.1) takes the form

$$F_i = -\frac{1}{R_{ij}^3} + \frac{0.06854}{R_{ij}^5}. \tag{10.3}$$

We choose F_i to be

$$F_i = E\left(-\frac{1}{R_{ij}^3} + \frac{0.06854}{R_{ij}^5}\right), \tag{10.4}$$

in which E is a positive constant called the constant of elasticity, for reasons to be discussed later.

Next note that the initial distance between a particle on C_1 and a neighbor on C_2 is 1.0205 cm, that for two neighbors on C_2 is 0.09865 cm, that for particle on C_2 and a neighbor on C_3 is 1.0288 cm, and that for two neighbors on C_3 is 0.1343 cm. Thus, just as (10.4) was derived, the magnitude of the force on P_i due to a neighbor P_j is given by

$$F_i = E\left(-\frac{1}{R_{ij}^3} + \frac{0.06854}{R_{ij}^5}\right), \quad P_i \text{ on } C_1, \ P_j \text{ on } C_1$$

$$F_i = E\left(-\frac{1}{R_{ij}^3} + \frac{1.0205}{R_{ij}^5}\right), \quad P_i \text{ on } C_1, \ P_j \text{ on } C_2$$

$$F_i = E\left(-\frac{1}{R_{ij}^3} + \frac{0.09865}{R_{ij}^5}\right), \quad P_i \text{ on } C_2, \ P_j \text{ on } C_2$$

$$F_i = E\left(-\frac{1}{R_{ij}^3} + \frac{1.0288}{R_{ij}^5}\right), \quad P_i \text{ on } C_2, \ P_j \text{ on } C_3$$

$$F_i = E\left(-\frac{1}{R_{ij}^3} + \frac{0.1343}{R_{ij}^5}\right), \quad P_i \text{ on } C_3, \ P_j \text{ on } C_3.$$

The forces just derived are local in that they act only between a particle and its neighbors. The force of gravity will be introduced and is long range in that it acts on all particles uniformly. The magnitude of gravity is denoted by g. Note that the equilibrium distance is unchanged by the choice of E. Thus we will consider a 360-body problem in which each particle is acted upon locally by its neighbors and also acted upon by gravity. From given initial data, the numerical solution will be generated by the leap frog formulas.

10.3. Examples

Throughout this section we will scale forces to reduce computational times. To accomplish this, we fix the time step to be 10^{-6}. In the examples we will consider the X axis to be a solid boundary. Initially the ball will be placed above the axis and we will require that it never fall below axis. However, numerical computations can yield a negative y_i for a particle P_i, and so a protocol is needed to handle this possibility. In such cases, if the point (x_i, y_i) has $y_i < 0$, then y_i will be replaced by $-y_i$, $v_{x,i}$ will be replaced by

$\alpha v_{x,i}$, and $v_{y,i}$ will be replaced by $\beta v_{y,i}$, $0 \leq \alpha \leq 1, -1 \leq \beta \leq 0$. The case $\alpha = 1, \beta = -1$ represents a frictionless interaction while the case $\alpha = 0, \beta = 0$ represents an interaction with full friction.

Example 1. Consider first the following simple example. Set $E = 1000, \alpha = 0, \beta = 0$, and scale g to be 0.098. This will provide a slow motion version of the dynamics. Next, raise the center of the ball in Figure 10.3 to be 10 cm above the X axis. Set all initial velocities equal to zero. This

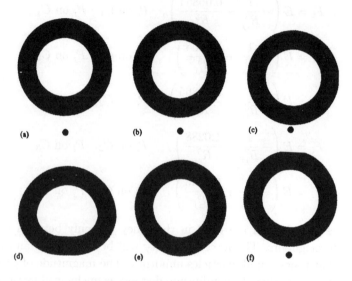

Figure 10.5. $E = 1000, g = 0.098$.

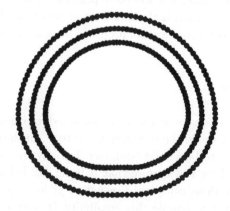

Figure 10.6.   Small radius form of Figure 10.5(d).

configuration is shown in Figure 10.5(a) at $t = 0$. Now let the ball drop from this position of rest. The resulting bouncing motion is shown every 3 million time steps, that is at $t = 0, 3, 6, 9, 12, 15$ in Figures 10.5(a)–10.5(f). Figure 10.5(d) shows that, for the present parameter choices, the ball shows extensive bending. Figure 10.6 shows Figure 10.5(d) using a smaller particle radius and reveals the mechanism of the bounce. In Figure 10.6, one sees that the flattened portion contains particle compression, which yields large repulsion between these particles, the culmination of which is the bounce effect in Figures 10.5(e) and 10.5(f).

Example 2. All the parameters of Example 1 were the same with the single exception $E = 16000$. Figures 10.7(a)–10.7(f) show the resulting bounce at the respective times $t = 0, 3, 6, 9, 12, 15$. Unlike Example 1, the ball now has greater local forces and is able to keep its relatively circular shape at all times.

Example 3. Example 1 was repeated with only two modifications. The center of the ball was raised 15 cm and the gravity constant was taken to be 0.98. The complete collapse of the ball is shown in Figures 10.8(a)–10.8(c) at the respective times $t = 0, 3, 6$. The local forces are no longer sufficient to counteract the resulting vertical force due to gravity and the result is a crash into the axis.

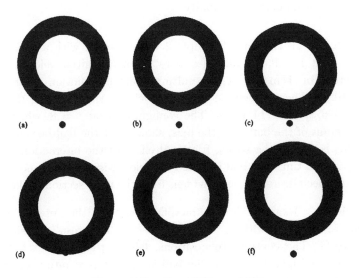

Figure 10.7. $E = 16000, g = 0.098$.

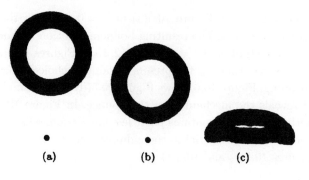

Figure 10.8. $E = 1000, g = 0.98.$

Example 4. Example 3 was rerun with $E = 16000$. As in Example 2, the local forces are greatly increased. The resulting bouncing is shown in Figures 10.9(a)–10.9(o) at the respective times $t = 0$, 1.5, 3.0, 4.5, 6.0, 7.5, 8.5, 9.0, 10.5, 12.0, 13.5, 15.0, 16.5, 18.0, 19.5. The behavior is now entirely similar to that in Example 1, and unlike that in Example 2. The resulting motion shows extensive bending.

From the Examples 1–4, we see that if one wishes to increase g, one need only increase E to have a ball which bounces, and which shows either a great deal of bending or very little bending. The computer problem in doing this is that as E increases, one must reduce the computational time step to prevent numerical instability.

Example 5. In this case Example 2 was repeated with a single modification, that is, all the particles were given an initial velocity in the X direction of $4\,\mathrm{cm/sec}$. The resulting bouncing motion is shown in Figure 10.10. The consecutive positions, left to right, are at the times $t = 0, 2, 4, 6, 8, 10, 12, 14, 16, 18$. The configuration at $t = 8$, which is the collision point of the ball with the axis, shows that the ball becomes irregular. This is due to the strong frictional effect of the interaction. Indeed, at this point in time, the ball is forced into clockwise rotation and its total horizontal speed decreases, as is evident from the right section of the figure.

Example 6. Example 5 was modified so that the interaction was frictionless, that is, with $\alpha = 1, \beta = -1$. The result is shown in Figure 10.11. The consecutive positions, left to right, are at the times $t = 0, 2, 4, 6, 8, 10, 12, 14, 16, 18$. The ball is never forced into rotation and never loses horizontal speed.

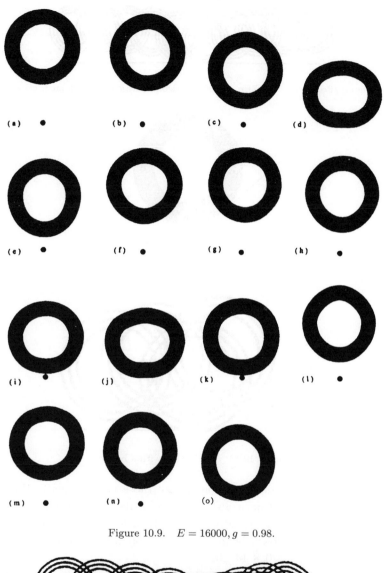

Figure 10.9. $E = 16000, g = 0.98$.

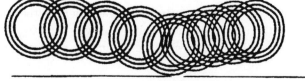

Figure 10.10. $E = 16000, g = 0.098$, horizontal trajectory with friction.

Figure 10.11. $E = 16000, g = 0.098$, horizontal trajectory without friction.

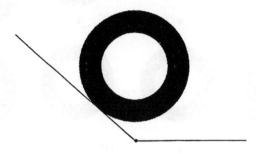

Figure 10.12.   Ball on a frictionless incline.

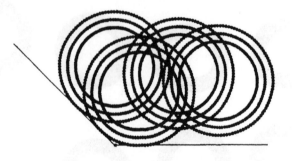

Figure 10.13.   Trajectory with a bounce.

Example 7.   The initial parameters are those of Example 2. However, the ball is now placed on a 45° incline with its center 10 cm above the X axis, as shown in Figure 10.12. In addition, the motion is taken to be frictionless. The roll down the incline and the bounce off the axis are shown in Figure 10.13 at $t = 0, 8, 14, 20$.

A variety of other examples were run in addition to those described above, usually with entirely similar results. This was the case for $r_1 = 10$, $r_2 = 11$, $r_3 = 12$, and for cases with the parameter choices $p = 2$, $q = 4$ in (10.1).

Chapter 11

Particle Model of String Solitons

11.1. Introduction

A soliton is a local, traveling wave pulse. It is significant in a variety of unrelated areas, like studies of water waves, acoustic waves, electron transfer, gravity waves, plasmas, optics and condensed matter (Miura (1976), Newell (1985)). In this chapter we will generate and study solitons in a new context, that is, from the point of view of discrete string motion in the XY plane. Our approach enables one to elucidate the complexities of soliton interactions easily using velocity profiles.

11.2. Discrete Strings

A discrete string is one which consists of $(N + 1)$ ordered particles $P_1, P_2, \ldots, P_{N+1}, N$ a positive integer, each of mass m. Assume that P_1 is fixed at $(0,0)$ and P_{N+1} is fixed at $(2,0)$, while the remaining $(N-1)$ particles can move. Assume also that each moving particle interacts only with its two neighbors. *Initially*, we allow the moving particles $P_2 - P_N$ to move only in the Y direction. The interval $0 \leq x \leq 2$ is now divided into N equal parts, each having grid size $\Delta x = 2/N$. A moving particle P_i is assigned the fixed x coordinate $(i-1)\Delta x$. For a given time step Δt, let $t_k = k\Delta t, k = 0, 1, 2, \ldots$. Denote the Y coordinate, the velocity, and the acceleration of P_i at t_k by $y_{i,k}$, $v_{i,k}$, and $a_{i,k}$, respectively. Let $|T_{i-1,i,k}|$ be the tensile force between P_{i-1} and P_i at t_k. Then, each $a_{i,k}$ is defined by the dynamical difference equation

$$
ma_{i,k} = |T_{i,i+1,k}| \frac{y_{i+1,k} - y_{i,k}}{[(\Delta x)^2 + (y_{i+1,k} - y_{i,k})^2]^{1/2}}
$$
$$
- |T_{i-1,i,k}| \frac{y_{i,k} - y_{i-1,k}}{[(\Delta x)^2 + (y_{i,k} - y_{i-1,k})^2]^{1/2}}, \quad i = 2, 3, \ldots, N. \quad (11.1)
$$

From (11.1), each $v_{i,k+1}$ is then determined from the special formulas (Greenspan (1972)):

$$\begin{cases} v_{i,1} = v_{i,0} + (\Delta t)a_{i,0} \\ v_{i,k+1} = v_{i,k} + (\Delta t)(1.5a_{i,k} - 0.5a_{i,k-1}), \quad k = 1, 2, 3, \ldots, \end{cases} \tag{11.2}$$

while, with the aid of (11.2), each $y_{i,k+1}$ is determined from

$$y_{i,k+1} = y_{i,k} + \frac{1}{2}(\Delta t)(v_{i,k+1} + v_{i,k}), \quad k = 0, 1, 2, \ldots. \tag{11.3}$$

From initial data $y_{i,0}, v_{i,0}, i = 2, 3, \ldots, N$, the motions of $P_2, P_3, \ldots, P_N$ are then determined recursively at each time step by (11.1)–(11.3).

It remains then to discuss the class of tension formulas to be explored in the computer examples which follow. In the present chapter these will be limited to tensile forces of the particular form

$$|T_{i,i+1,k}| = T_0 \left[(1 - \epsilon) \left(\frac{r_{i,i+1,k}}{\Delta x} \right) + \epsilon \left(\frac{r_{i,i+1,k}}{\Delta x} \right)^2 \right] \tag{11.4}$$

in which $0 \leq \epsilon \leq 1, T_0$ is a reference tension, and

$$r_{i,i+1,k} = [(\Delta x)^2 + (y_{i+1,k} - y_{i,k})^2]^{1/2}. \tag{11.5}$$

Note that if all the particles are on the X axis, then

$$|T_{i,i+1,k}| = T_0, \quad i = 1, 2, 3, \ldots, 1000. \tag{11.6}$$

11.3.   Examples

In all the examples, the parameters are $N = 1000, m = 0.01, T_0 = 10$, and, unless otherwise specified, $\Delta x = 0.002$, $\epsilon = 0.0$ and $\Delta t = 0.0001$. Thus the fixed particles are P_0 and P_{1001}, while the moving particles are $P_2 - P_{1000}$. The initial positions of the moving particles are all set on the X axis, so that in each example we need only specify their initial velocities.

Example 1. A soliton is created on the left by assigning all initial velocities equal to zero except at $P_2 - P_{63}$, and these are assigned as follows:

$$v_{i,0}(P_2) = 0.5$$
$$v_{i,0}(P_{i+1}) = v_{i,0}(P_i) + 0.5, \qquad i = 2, 31$$
$$v_{i,0}(P_{33}) = v_{i,0}(P_{32}) = 15.5$$
$$v_{i,0}(P_{i+1}) = v_{i,0}(P_i) - 0.5, \qquad i = 33, 62.$$

The resulting soliton motion is shown in Figure 11.1 at the times 0.15, 0.45, 0.75, and 1.05. The speed of the soliton is constant at 0.707 cm/sec and, to three significant digits, its amplitude is constant at 0.351 cm. To three significant digits the kinetic energy is 1300 at all times. What is not apparent in the figure are the small trailing waves which follow behind the soliton. A portion of these are shown in Figure 11.2 in an enlarged section of the motion at the time 0.75 and with the use of a smaller particle radius. The amplitude of the trailing wave nearest the soliton is −0.001 cm. The amplitudes of the previous trailing waves are as small as ±0.000001. Trailing waves appear in all the examples which follow.

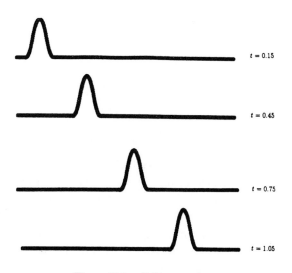

$t = 0.15$

$t = 0.45$

$t = 0.75$

$t = 1.05$

Figure 11.1. Soliton motion.

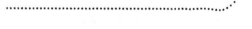

Figure 11.2.   Trailing waves.

Example 2.   A soliton on the left is generated as in Example 1. In addition, a smaller soliton is generated on the right as follows:

$$v_{i,0}(P_{1000}) = 0.5$$

$$v_{i,0}(P_{999}) = 1.0$$

$$v_{i,0}(P_{999-i}) = v_{i,0}(P_{1000-i}) + 0.5, \quad i = 1, 23$$

$$v_{i,0}(P_{975}) = v_{i,0}(P_{976}) = 12.5$$

$$v_{i,0}(P_{975-i}) = v_{i,0}(P_{976-i}) - 0.5, \quad i = 1, 24.$$

Figure 11.3 shows clearly at the times 0.15, 0.60, 0.75, 0.90, 1.05 that the collision of the two solitons results in their passing through each other. Figure 11.4 shows in greater detail the soliton interactions at the times 0.635, 0.650, 0.650, 0.680, 0.695, 0.710, 0.725, 0.740, 0.755, 0.770, 0.785. The complex interaction shown in Figure 11.4 is more clearly delineated in Figure 11.5, which shows the velocities of each of the particles at the very same times as those in Figure 11.4.

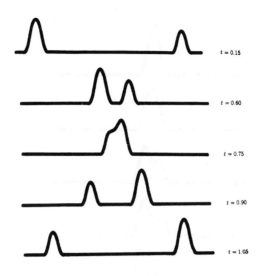

Figure 11.3.   Soliton collision.

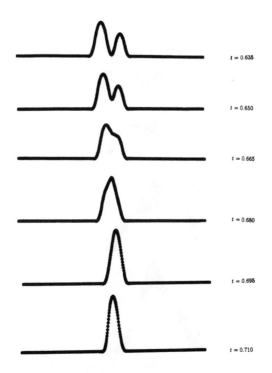

Figure 11.4.   Detailed soliton collision.

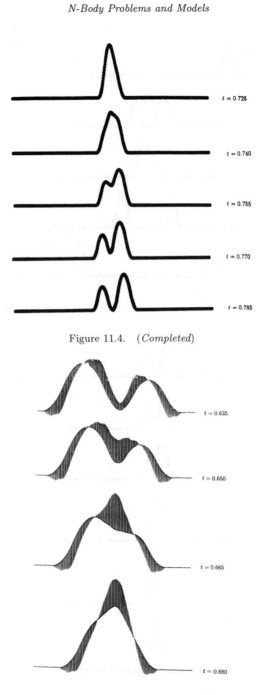

Figure 11.4. (*Completed*)

Figure 11.5. Soliton velocity profiles.

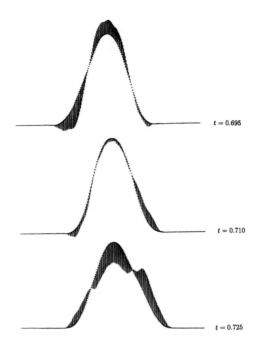

Figure 11.5. (*Continued*)

Example 3. Example 2 was modified by setting

$$v_{i,0}(P_{1000}) = -0.5$$
$$v_{i,0}(P_{999}) = -1.0$$
$$v_{i,0}(P_{999-i}) = v_{i,0}(P_{1000-i}) - 0.5, \quad i = 1, 23$$
$$v_{i,0}(P_{975}) = v_{i,0}(P_{976}) = -12.5$$
$$v_{i,0}(P_{975-i}) = v_{i,0}(P_{976-i}) + 0.5, \quad i = 1, 24.$$

Figure 11.6 shows clearly at the times 0.15, 0.60, 0.75, 0.90, 1.05 that the collision of the two solitons results in their passing through each other even though their amplitudes are of different signs.

Example 4. Figure 11.7 shows the motion at the times 0.15, 0.45, 0.75, 1.05 when the soliton parameters for Example 1 contain the single change $\epsilon = 0.02$. The soliton's motion shows a bending behavior with time and a sharpening of the apex. The motion at time 1.05 is shown twice, first with the particle radius usually used and then with a decrease in this radius to elucidate the contractive bunching of particles within the soliton.

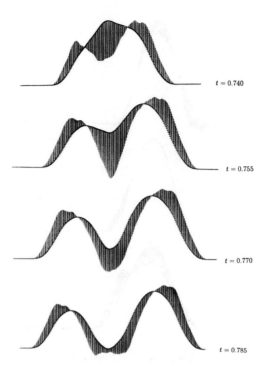

Figure 11.5.   (*Completed*)

Example 5.   Figure 11.8 shows the motion at the times 0.15, 0.45, 0.75, 1.05 of the soliton generated in Example 1 when motion in the X direction is included in the model. The figure appears to be essentially identical with that of Figure 11.1, and this was verified by comparing the computer outputs.

Example 6.   Example 2 was repeated with motion in the X direction included and the results were essentially identical to those of Example 2.

Example 7.   Example 1 was repeated allowing for motion in the X direction and setting $\epsilon = 0.02$. The soliton became physically unstable at the time 0.45 and fell apart. The same result followed for the time steps $\Delta t = 0.00001, 0.000001$. We then reset $\Delta t = 0.0001$ and studied $\epsilon = 0.01$, 0.001, 0.0001. The effect of decreasing ϵ was to decrease the amplitude of the soliton, as is shown in Figure 11.9 at the times 0.15, 0.45, 0.75 for $\epsilon = 0.01$. This effect decreases as ϵ decreases.

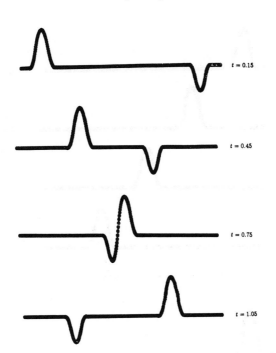

Figure 11.6.   Soliton interaction.

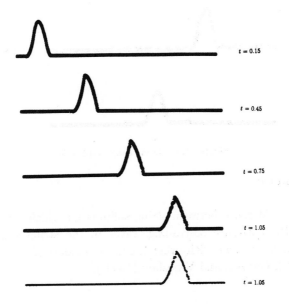

Figure 11.7.   Nonzero epsilon.

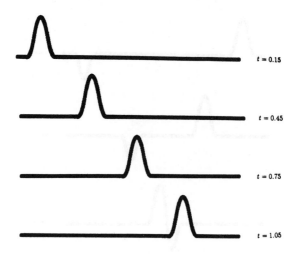

Figure 11.8.   Nonzero horizontal component.

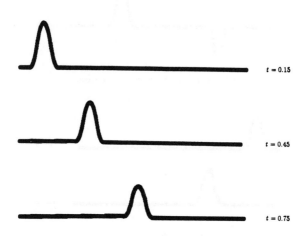

Figure 11.9.   Decreasing amplitude.

11.4.   Remark

The primary difference between string solitons for which $\epsilon = 0$ and KdV solitons is the existence of trailing waves in the string solitons, which are, however, consistent with other nonlinear models of vibrating strings (Ames (1969), Gotusso and Veneziani (1994)).

Chapter 12

Particle Models of Minimal Surfaces and Saddle Surfaces

12.1. Introduction

Soap films and minimal surfaces have long been of interest both mathematically and physically (Almgren and Taylor (1976), Courant (1950), Rado (1960)). The Belgian physicist J. A. F. Plateau (1801–1883) determined experimentally a number of geometric properties of soap films by dipping a closed, thin wire into a soap solution and studying the resulting soap film, or minimal surface, which spanned the wire. Plateau concluded that every closed wire contour always bounded a minimal surface. Nevertheless, the nonuniqueness of the problem is now well known (Courant (1950)).

In the next section we will model minimal surfaces using the particle approach.

12.2. A Particle Model of a Minimal Surface

Consider a class of problems in which the boundary curves are three dimensional and are of the type shown in Figure 12.1. C_1, C_2 are semicircular, parallel and have equal radii. L_1, L_2 are parallel straight line segments of equal length. The planes of C_1, C_2 are perpendicular to L_1, L_2.

Consider then first a rectangular array of 43 particles P_1-P_{43} shown in Figure 12.2. Let each particle be of unit mass. The distance between any two adjacent particles in the horizontal direction is taken to be 1.00645, while the distance between any two adjacent rows of particles is taken to be 0.866. The particles $P_{10}-P_{17}$, $P_{19}-P_{25}, P_{27}-P_{34}$, are called the interior particles while the others are called the boundary particles. By construction, interior particles $P_{10}, P_{17}, P_{27}, P_{34}$ have five neighbors each, while the remaining interior particles have six neighbors each. The arrangement

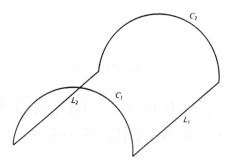

Figure 12.1.   A three dimensional wire.

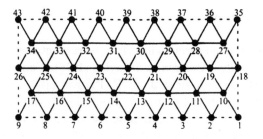

Figure 12.2.   Plane configuration.

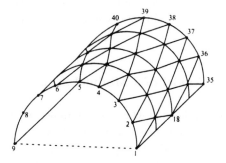

Figure 12.3.   Folded configuration.

shown in Figure 12.2 is then folded into a right cylindrical surface, as
shown in Figure 12.3. The particles P_1-P_9 are on C_1; $P_{35}-P_{43}$ are on
C_2; P_1, P_{18}, P_{35} are on L_1; and P_9, P_{26}, P_{43} are on L_2. The radii of both
C_1, C_2 are 2.56292 and the distance between any two interior particles and
each of its neighbors is unity.

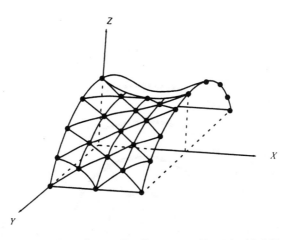

Figure 12.4. Saddle surface. (From: D. Greenspan, *Particle Modelling*, Birkhauser, Boston, 1997, p. 46.)

The force between any two particles is taken to have magnitude F given by

$$F_{ij} = -\frac{1}{r_{ij}^2} + \frac{0.5}{r_{ij}^4}.$$

Gravity is neglected. In applying the leap frog formulas to the resulting dynamical equations, the time step is $\Delta t = 0.001$. The boundary particles are fixed for all time and the interior particles are assigned zero initial velocities. Thereafter, the interior particles are allowed to vibrate to steady state, with convergence hastened by resetting all velocities equal to zero every 10,000 time steps. Each particle has interaction only with its neighbors. The steady state configuration is shown in Figure 12.4. For precise details see Greenspan (1997), Chapter 5.

12.3. A Monkey Saddle

Figure 12.4 is also an example of a saddle surface. A mathematical saddle surface is the graph of a hyperbolic paraboloid whose equation is

$$\frac{x^2}{a^2} - \frac{y^2}{b^2} = cz \quad (a \neq 0, b \neq 0, c \neq 0).$$

The surface has the general shape of a saddle with two upward slopes and, for the legs, two downward slopes. Another type of saddle surface is a monkey saddle, which has three downward slopes, two for the legs and one

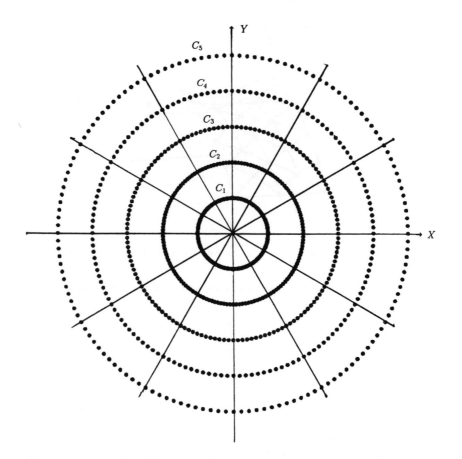

Figure 12.5.   Concentric particles.

for the tail. However, no simple formula like the one above exists for the monkey saddle.

In this section we will show how to generate a monkey saddle. The generation of a monkey saddle is more complex than that described in Section 12.2. In addition, for variety, we will choose forces this time with magnitudes given by

$$F_{ij} = -\frac{A}{R_{ij}^1} + \frac{B}{R_{ij}^3}, \quad A > 0, \ B > 0, \tag{12.1}$$

We begin by considering a fixed point P_1 at the origin and 720 additional particles P_2-P_{721}, arranged on 5 concentric circles, C_1-C_5, as shown in Figure 12.5. The distance between any two adjacent circles is chosen to be

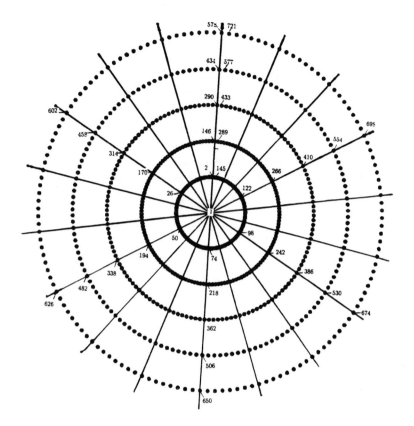

Figure 12.6.   Particle numbering.

unity. On each circle there are 144 particles determined by intersections with the rays which emanate from the the origin at the angles $(2.5k)°$, $k = 0, 1, 2, \ldots, 143$, relative to the positive Y axis. The particles are numbered 2–721, consecutively, counterclockwise from the positive Y axis, beginning on the circle C_1, continuing to circle C_2, then to circle C_3, then to C_4 and finally to C_5. Typical and important particle numbers are shown in Figure 12.6.

The neighbors of any particle on C_1 are P_1, the two adjacent particles on C_1, and the closest particle on C_2. Thus, each particle on C_1 has exactly four neighbors. The neighbors of any particle on $C_2 - C_4$ are the two particles nearest to it on the same circle and the nearest particle on each of the adjacent circles. Thus, each of the particles $P_{146} - P_{577}$ also has exactly four neighbors. Finally, for any particle on the circle C_5, the neighbors are defined to be the two adjacent particles on C_5 and the closest particle on C_4. Thus, any particle on C_5 has exactly three neighbors. The neighbors of any particle are its neighbors for all time.

Dynamically, each particle will interact only with its neighbors. However, it is desirable that the force between any two particles in Figure 12.6 be zero initially, so that the particles are initially in equilibrium. But, since the distance between two neighboring particles on the same circle varies from circle to circle, a single force formula does not suffice, so that we define force magnitudes in the following way.

If P_i and P_j are neighbors, but are on adjacent circles, then let

$$F_{ij} = \frac{1000}{R_{ij}^3} - \frac{1000}{R_{ij}^1}.$$
(12.2)

If P_i and P_j are neighbors and are adjacent on C_1, let

$$F_{ij} = \frac{1.900}{R_{ij}^3} - \frac{1000}{R_{ij}^1}.$$
(12.3)

If P_i and P_j are neighbors and are adjacent on C_2, let

$$F_{ij} = \frac{7.610}{R_{ij}^3} - \frac{1000}{R_{ij}^1}.$$
(12.4)

If P_i and P_j are neighbors and are adjacent on C_3, let

$$F_{ij} = \frac{17.13}{R_{ij}^3} - \frac{1000}{R_{ij}^1}.$$
(12.5)

If P_i and P_j are neighbors and are adjacent on C_4, let

$$F_{ij} = \frac{30.46}{R_{ij}^3} - \frac{1000}{R_{ij}^1}.$$
(12.6)

Finally, if P_i and P_j are neighbors and are adjacent on C_5, let

$$F_{ij} = \frac{45.68}{R_{ij}^3} - \frac{1000}{R_{ij}^1}.$$
(12.7)

The use of formulas (12.2)–(12.7) results in the equilibrium of Figure 12.6, since the distance between two neighbors on C_1 is 0.043629, between two neighbors on C_2 is 0.087260, between two neighbors on C_3 is 0.130888, between two neighbors on C_4 is 0.174520, and between two neighbors on C_5 is 0.213742.

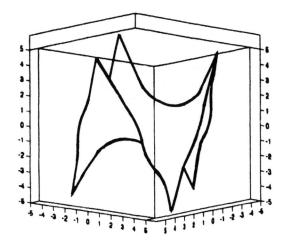

Figure 12.7. Monkey saddle.

Denoting the z coordinate of P_i by $z(i)$, let $z(i) = 0$ initially for all the particles in Figure 12.6. We now disturb the equilibrium of Figure 12.6 by creating alternating upward and downward slopes, or spines, every 60° as follows.

$$z(578) = z(626) = z(674) = -z(602) = -z(650) = -z(698)$$
$$= 5 \tag{12.8}$$
$$z(434) = z(482) = z(530) = -z(458) = -z(506) = -z(554)$$
$$= 2.434 \tag{12.9}$$
$$z(290) = z(338) = z(386) = -z(314) = -z(362) = -z(410)$$
$$= 1.08. \tag{12.10}$$
$$z(146) = z(194) = z(242) = -z(170) = -z(218) = -z(266)$$
$$= 0.32 \tag{12.11}$$
$$z(2) = z(50) = z(98) = -z(26) = -z(74) = -z(122)$$
$$= 0.04 \tag{12.12}$$
$$z(1) = 0. \tag{12.13}$$

The points at which these values are prescribed will remain fixed throughout the dynamical considerations.

We now consider a system of 690, nonlinear, vector second order Newtonian dynamical equations, one for each particle which has not been fixed. Each such particle interacts only with its neighbors through (12.2)–(12.7). Gravity is neglected. The moving particles are assigned zero initial velocities. The dynamical system is solved numerically with the leap frog formulas using the time step $\Delta t = 0.00002$ and the velocities are damped periodically to ease the configuration into steady state. The resulting steady state monkey saddle is shown in Figure 12.7.

The fixed values in (12.8)–(12.13) were determined by substituting the y coordinates of $P_{578}, P_{434}, P_{290}, P_{146}, P_2$ and P_1 into the function $y^3/125$. Similar choices led to completely analogous monkey saddles.

Appendix I

A Generic Program for Kutta's Fourth Order Formulas for Second Order Initial Value Problems

Step 1.   Set a counter $k = 1$.

Step 2.   Set a time step h.

Step 3.   Set an initial time x.

Step 4.   Set initial values y, v.

Step 5.   Calculate
$$M_0 = hf(x, y, v)$$
$$M_1 = hf(x + \tfrac{1}{2}h, y + \tfrac{1}{2}hv, v + \tfrac{1}{2}M_0)$$
$$M_2 = hf(x + \tfrac{1}{2}h, y + \tfrac{1}{2}hv + \tfrac{1}{4}hM_0, v + \tfrac{1}{2}M_1)$$
$$M_3 = hf(x + h, y + hv + \tfrac{1}{2}hM_1, v + M_2)$$

Step 6.   Calculate y at $x + h$ and v at $x + h$ by
$$y(x + h) = y + hv + \tfrac{1}{6}h(M_0 + M_1 + M_2)$$
$$v(x + h) = v + \tfrac{1}{6}(M_0 + 2M_1 + 2M_2 + M_3)$$

Step 7.   Increase the counter from k to $k + 1$.

Step 8.   Set $y = y(x + h), v = v(x + h), x = x + h$

Step 9.   Repeat Steps 5–8.

Step 10.   Continue until $k = 100$.

Step 11.   Stop the calculation.

A Generic Program for Kutta's Fourth Order Formulas for Second Order Initial Value Problems

Appendix II

Newton's Iteration Formulas
for Systems of Algebraic
and Transcendental Equations

In order to solve a system of k algebraic or transcendental equations in the k unknowns $x_1, x_2, \ldots, x_k$, that is,

$$f_1(x_1, x_2, \ldots, x_k) = 0$$
$$f_2(x_1, x_2, \ldots, x_k) = 0$$
$$\vdots$$
$$f_k(x_1, x_2, \ldots, x_k) = 0,$$

iterate the Newtonian formulas to convergence from an initial guess with

$$x_1^{(n+1)} = x_1^{(n)} - \frac{f_1(x_1^{(n)}, x_2^{(n)}, \ldots, x_k^{(n)})}{\frac{\partial f_1}{\partial x_1}(x_1^{(n)}, x_2^{(n)}, \ldots, x_k^{(n)})}$$

$$x_2^{(n+1)} = x_2^{(n)} - \frac{f_2(x_1^{(n)}, x_2^{(n)}, \ldots, x_k^{(n)})}{\frac{\partial f_2}{\partial x_2}(x_1^{(n)}, x_2^{(n)}, \ldots, x_k^{(n)})}$$

$$\vdots$$

$$x_k^{(n+1)} = x_k^{(n)} - \frac{f_k(x_1^{(n)}, x_2^{(n)}, \ldots, x_k^{(n)})}{\frac{\partial f_k}{\partial x_k}(x_1^{(n)}, x_2^{(n)}, \ldots, x_k^{(n)})}.$$

Often the initial guess $x_1^{(0)} = x_2^{(0)} = \ldots = x_k^{(0)} = 0.0$ is adequate. However, if convergence does not result, then a different initial guess must be chosen. Since no proof has been provided that a solution of the equations exists, one must provide an *a posteriori* proof by substitution of the result of the convergent iteration into the original set of equations to show that it is a solution.

Appendix III

The Leap Frog Formulas

Choose a positive time step Δt and let $t_k = k\Delta t, k = 0, 1, 2, \ldots$. For $i = 1, 2, 3, \ldots, N$, let P_i have mass m_i and at t_k let it be at

$$\vec{r}_{i,k} = (x_{i,k}, y_{i,k}, z_{i,k});$$

have velocity

$$\vec{v}_{i,k} = (v_{i,k,x}, v_{i,k,y}, v_{i,k,z}),$$

and have acceleration

$$\vec{a}_{i,k} = (a_{i,k,x}, a_{i,k,y}, a_{i,k,z}).$$

The leap frog formulas, which relate position, velocity and acceleration are

$$\frac{\vec{v}_{i,1/2} - \vec{v}_{i,0}}{(1/2)\Delta t} = \vec{a}_{i,0}, \qquad \text{(Starter)} \tag{1.10}$$

$$\frac{\vec{v}_{i,k+1/2} - \vec{v}_{i,k-1/2}}{\Delta t} = \vec{a}_{i,k}, \qquad k = 1, 2, 3, \ldots, \tag{1.11}$$

$$\frac{\vec{r}_{i,k+1} - \vec{r}_{i,k}}{\Delta t} = \vec{v}_{i,k+1/2}, \qquad k = 0, 1, 2, 3, \ldots, \tag{1.12}$$

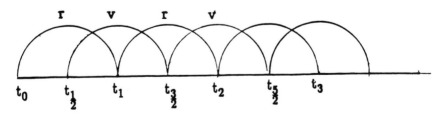

Figure A3.1. Leap frog.

or, explicitly,

$$\vec{v}_{i,\frac{1}{2}} = \vec{v}_{i,0} + \frac{1}{2}(\Delta t)\vec{a}_{i,0}, \quad \text{(Starter)} \tag{1.13}$$

$$\vec{v}_{i,k+\frac{1}{2}} = \vec{v}_{i,k-\frac{1}{2}} + (\Delta t)\vec{a}_{i,k}, \quad k = 1, 2, 3, \ldots, \tag{1.14}$$

$$\vec{r}_{i,k+1} = \vec{r}_{i,k} + (\Delta t)\vec{v}_{i,k+\frac{1}{2}}, \quad k = 0, 1, 2, 3 \ldots. \tag{1.15}$$

Note that (1.11) and (1.12) are two point central difference formulas. The name *leap frog* derives from the way position and velocity are defined at alternate, sequential time values . As shown in the Figure A3.1, the r values are defined at the times $t_0, t_1, t_2, t_3, \ldots$, while the v values are defined at the times $t_{\frac{1}{2}}, t_{\frac{3}{2}}, t_{\frac{5}{2}}, t_{\frac{7}{2}}, \ldots$. The figure also symbolizes the children's game "leap frog".

References and Additional Sources

Adam, J. R., N. R. Lindblad and C. D. Hendricks, "The collision, coalescence and disruption of water droplets", *J. Appl. Phys.*, 39, 1968, p. 5173.

Adamson, A. W., *Physical Chemistry of Surfaces*, Interscience, N.Y., 1976.

Almgren, F. J. and J. E. Taylor, "The geometry of soap films and soap bubbles", *Sci. Amer.*, 235, 1976, p. 82.

Ames, W. F., *Numerical Methods for Partial Differential Equations*, Barnes and Noble, N.Y., 1969, pp. 201–202.

Ashurst, W. T. and W. G. Hoover, "Microscopic fracture studies in the two-dimensional triangular lattice", *Phys. Rev.*, B14, 1976, p. 1465.

Bachelor, G. K., *The Theory of Homogeneous Turbulence*, Cambridge University Press, Cambridge, 1959.

Bergmann, P. G., *Introduction to the Theory of Relativity*, Dover, N.Y., 1976, Chapter IV.

Bernard, P. S., "Transition and turbulence. Basic physics", in *The Handbook of Fluid Dynamics*, CRC Press, Boca Raton, 1998, pp. 13–15.

Calogero, F., "Exactly solvable one-dimensional many body problems", *Lett. Nuovo Cimento*, 11, 1975, pp. 411–416.

Coppin, C. and D. Greenspan, "A contribution to the particle modelling of minimal surfaces", *Appl. Math. Comp.*, 13, 1983, p. 17.

Courant, R., *Dirichlet's Principle, Conformal Mapping, and Minimal Surfaces*, Interscience, N.Y., 1950.

Feynman, R. P., R. B. Leighton and M. Sands, *The Feynman Lectures on Physics*, Vol. I, Addison-Wesley, Reading, 1963.

Freitas, C. J., R. L. Street, A. N. Findikakis and J. R. Koseff, "Numerical simulation of three dimensional flow in a cavity", *Int. J. Num. Meth. Fluids*, 5, 1985, pp. 561–575.

Girifalco, L. A. and R. A. Lad, "Energy cohesion, compressibility, and the potential energy functions of the graphite system", *J. Chem. Phys.*, 25, 1956, 693.

Goldstein, H., *Classical Mechanics*, 2nd ed., Addison-Wesley, Reading, 1980, pp. 54–63.

Gotusso, L. and A. Veneziani, "Discrete and continuous nonlinear models for the vibrating string", Tech. Rpt. n.143/P, Dip. Mat., Politecnico di Milano, 1994.

Greenspan, D., "A new explicit discrete mechanics with applications", *J. Franklin Inst.*, 294, 1972, pp. 231–240.

Greenspan, D., *Arithmetic Applied Mathematics*, Pergamon, Oxford, 1980.

Greenspan, D., "Particle simulation of biological sorting", TR 254, Math. Dept., Univ. Texas at Arlington, 1988.

Greenspan, D., "Particle simulation of biological sorting on a supercomputer", *Comp. Math. Applic.*, 18, 1989, pp. 823–834.

Greenspan, D., *Particle Modelling*, Birkhauser, Boston, 1997.

Griffiths, D. J., *Introduction to Electrodynamics*, Prentice Hall, Englewood Cliffs, 1981.

Hirschfelder, J. O., C. F. Curtiss, and R. B. Bird, *Molecular Theory of Gases and Liquids*, Wiley, N.Y., 1967.

Johnson, K. L., *Contact Mechanics*, Cambridge Univ. Press, Cambridge, 1985.

Keller, H. B. and E. L. Reiss, "Spherical cap snapping", *J. Aero/Space Sci.*, 26, 1959, p. 643.

Kelly, B. T., *Physics of Graphite*, Applied Sci., London, 1981.

Klabring, A., "On discrete and discretized nonlinear elastic structures in unilateral contact", in. *J. Solids and Structures*, 24, 1988, pp. 4–8.

Kolmogorov, A. N., "Toward a more precise notion of the structure of the local turbulence in a viscous fluid at elevated Reynolds number" in *The Mechanics of Turbulence* (A. Favre, Ed.), Gordon and Breach, New York, 1964, pp. 447–458.

Korlie, M., *Particle Modeling of a Liquid Drop on a Solid Surface in 3-D*, PhD Thesis, Math., UT Arlington, 1996.

Ladyzhenskaya, O. A., *The Mathematical Theory of Viscous Incompressible Flow*, 2nd ed., Gordon and Breach, N.Y., 1969.

Lebon, F. and M. Raous, "Multibody contact problems including friction in structure assembly", *Computers and Structures*, 43, 1992, pp. 925–934.

Marsden, J. E., *Lectures on Geometric Methods in Mathematical Physics*, SIAM, Philadelphia, 1981, p. 39.

Martins, J. A. C. and J. T. Oden, "Existence and uniqueness results for dynamic contact problems with nonlinear normal and friction interface laws", *Nonlinear Analysis*, 11, 1987, pp. 407–428.

Megaw, H. D., *Crystal Structure: A Working Approach*, W. B. Saunders, Phila., 1973.

Miura, R., "The Korteweg-de Vries equation, a survey of results", *SIAM Rev.*, 18, 1976, pp. 412–459.

Newell, A. C., *Solitons in Mathematics and Physics*, CBMS 48, SIAM, 1985.

Pan, F. and A. Acrivos, "Steady flows in a rectangular cavity", *J. Fluid Mech.*, 28, 1967, pp. 643–655.

Peterson, I., "Raindrop oscillation", *Sci. News*, 2, 1985, p. 136.

Raous, M., P. Chabrand and F. Lebob, "Numerical methods and frictional contact problems and applications", *J. Theor. Appl. Mech.*, 7, 1988, pp. 111–128.

Rado, T., *On the Plateau Problem*, Chelsea, N.Y., 1951.

Schlichting, H., *Boundary Layer Theory*, 4th ed., Mcgraw-Hill, New York, 1960.

Sears, F. W. and M. W. Zemansky, *University Physics*, 2nd ed., p. 315, Addison-Wesley, Reading, 1957.

Shillor, M., Ed., *Recent Advances in Contact Mechanics*, Special Issue of *Mathl. Comput. Modelling*, 287, 1988.

Simpson, S. F. and F. J. Haller, "Effects of experimental variables on mixing of solutions by collision of microdroplets", *Analyt. Chem.*, 60, 1988, p. 2483.

Steinberg, M. S., "Reconstruction of tissues by dissociated cells", in *Models for Cell Rearrangement*, G. D. Mostow (Ed.), Yale University Press, New Haven, 1975, pp. 82–99.

Synge, J. L. and B. A. Griffith, *Principles of Mechanics*, McGraw-Hill, New York, 1942, Chapter VI.

Toda, M., "Wave propagation in an harmonic lattice", *J. Phys. Soc. Japan*, 23, 1967, pp. 501–506.

van de Kamp, P., *Elements of Astromechanics*, Freeman, San Francisco, 1964, Chapter III.

Index